Springer Theses

Recognizing Outstanding Ph.D. Research

Aims and Scope

The series "Springer Theses" brings together a selection of the very best Ph.D. theses from around the world and across the physical sciences. Nominated and endorsed by two recognized specialists, each published volume has been selected for its scientific excellence and the high impact of its contents for the pertinent field of research. For greater accessibility to non-specialists, the published versions include an extended introduction, as well as a foreword by the student's supervisor explaining the special relevance of the work for the field. As a whole, the series will provide a valuable resource both for newcomers to the research fields described, and for other scientists seeking detailed background information on special questions. Finally, it provides an accredited documentation of the valuable contributions made by today's younger generation of scientists.

Theses are accepted into the series by invited nomination only and must fulfill all of the following criteria

- They must be written in good English.
- The topic should fall within the confines of Chemistry, Physics, Earth Sciences, Engineering and related interdisciplinary fields such as Materials, Nanoscience, Chemical Engineering, Complex Systems and Biophysics.
- The work reported in the thesis must represent a significant scientific advance.
- If the thesis includes previously published material, permission to reproduce this must be gained from the respective copyright holder.
- They must have been examined and passed during the 12 months prior to nomination.
- Each thesis should include a foreword by the supervisor outlining the significance of its content.
- The theses should have a clearly defined structure including an introduction accessible to scientists not expert in that particular field.

More information about this series at http://www.springer.com/series/8790

Diana Bachiller Perea

Ion-Irradiation-Induced Damage in Nuclear Materials

Case Study of a-SiO$_2$ and MgO

Doctoral Thesis accepted by
the Université Paris-Sud, Orsay, France and
the Universidad Autónoma de Madrid, Madrid, Spain

Springer

Author
Dr. Diana Bachiller Perea
Detector Group
SOLEIL Synchrotron
Saint-Aubin, France

Supervisors
Dr. Aurélien Debelle
Centre de Sciences Nucléaires et de Sciences
  de la Matière (CSNSM)
Université Paris-Sud/CNRS-IN2P3/
  Université Paris-Saclay
Orsay, France

Dr. Ángel Muñoz Martín
Edificio de Rectorado
Universidad Autónoma de Madrid
Madrid, Spain

Dr. David Jiménez Rey
Laboratorio Nacional de Fusión
Centro de Investigaciones Energéticas,
  Medioambientales y Tecnológicas
Madrid, Spain

ISSN 2190-5053   ISSN 2190-5061   (electronic)
Springer Theses
ISBN 978-3-030-13111-1   ISBN 978-3-030-00407-1   (eBook)
https://doi.org/10.1007/978-3-030-00407-1

This Springer imprint is published by the registered company Springer Nature Switzerland AG
The registered company address is: Gewerbestrasse 11, 6330 Cham, Switzerland

To my parents

Supervisors' Foreword

The development of an energy source to supply the world's expanding needs with a limited environmental detriment has definitely become one of the most important current challenges for scientists. For this purpose, a huge international research effort is being devoted to the study of new systems of nuclear energy production, which includes Generation IV nuclear fission reactors and fusion reactors such as ITER. Among various technological locks that must be undone, materials represent a key issue that is tackled through numerous research programs worldwide. One major concern is related to the behavior of materials under the extreme environments that will be encountered in these new nuclear reactors, in particular the high levels of radiation by different highly energetic particles. Therefore, improving the resistance of already existing materials or developing new materials is of upmost importance. To achieve this goal, since the reactors are not yet built, it is mandatory to develop strategies and tools to study the materials' behavior under radiation environments that aim at emulating the foreseeable conditions of use.

One way to undertake such a task is to conduct fundamental studies on materials with identified potential applications using external ion beams delivered by particle accelerators. The advantage of this approach to emulate radiation environments lies in the possibility to vary the irradiation conditions (flux, fluence, temperature, ion nature, and energy) in a controlled way. Furthermore, the use of advanced, powerful techniques to characterize the materials allows determining the key parameters and the driving forces involved in the basic mechanisms of the generation of radiation-induced effects.

The thesis work carried out by Diana Bachiller Perea falls within this context and contributes to the progress in this major stake. Two aspects have been dealt with: the development of the use of a new technique to characterize radiation effects in materials, namely, the Ion Beam Induced Luminescence (IBIL) or ionoluminescence, and the study of two promising materials, amorphous silica (a-SiO_2) and (crystalline) magnesium oxide (MgO). IBIL was used on three different types of silica (containing different amounts of OH groups). Measurements were carried out at several ion beam facilities worldwide: at Universidad Autónoma de Madrid (Spain), at University of Knoxville (USA), and at Université Paris-Sud (France).

Results allowed to characterize the radiation effects, from the atomic scale by identifying the point defects created under irradiation, up to the macroscopic scale by monitoring the silica compaction. A phenomenological model was developed to reproduce the experimental data taking into account these two different effects. Radiation effects in MgO have been characterized using Rutherford backscattering spectrometry in channeling configuration and high-resolution X-ray diffraction. Results allowed proposing a detailed scenario of the microstructural changes occurring in this material under irradiation at elevated temperatures.

This thesis shall certainly be of interest for both beginners and experts alike in the field of radiation effects in materials. Indeed, in addition to the results mentioned above, which were obtained through a vast array of experiments, the book presents in a pedagogical and accessible way the basics of the experimental techniques used, as well as a detailed description of the context of the study and of the investigated materials.

To finish, it is time to introduce Diana Bachiller Perea. Diana started her Ph.D. at Universidad Autónoma de Madrid where she spent 2 years exploring and developing the IBIL technique. Then, thanks to the Eiffel Grants for Excellence program, she moved to Université Paris-Sud where she conducted in parallel the continuation of her work started in Madrid and her new study on magnesium oxide. The quality of her work, and thus her personal skills and knowledge, was recognized during and after her Ph.D. by several prestigious awards she received, in particular by the Award to the Best Ph.D. Thesis in Paris-Saclay University in the category Technology, Energy and Health awarded by the Laboratoire d'Excellence de Physique des Deux Infinis et des Origins (Labex P2IO) and the Award to the Best Ph.D. Thesis in Spain on Technological and Experimental Sciences awarded by the Royal Academy of Doctors of Spain (Real Academia de doctores de España). The publication by Springer of her thesis manuscript represents another recognition of that quality and we, supervisors, are pleased for contributing to Diana's scientific and personal "success-story".

Orsay, France Dr. Aurélien Debelle
Madrid, Spain Dr. David Jiménez Rey
Madrid, Spain Dr. Ángel Muñoz Martín
August 2018

Abstract

One of the most important challenges in Physics today is the development of a clean, sustainable, and efficient energy source that can satisfy the needs of the actual and future society producing the minimum impact on the environment. For this purpose, a huge international research effort is being devoted to the study of new systems of energy production; in particular, Generation IV fission reactors and nuclear fusion reactors are being developed. The materials used in these reactors will be subjected to high levels of radiation, making necessary the study of their behavior under irradiation to achieve a successful development of these new technologies.

In this thesis, two materials have been studied: amorphous silica (a-SiO$_2$) and magnesium oxide (MgO). Both materials are insulating oxides with applications in the nuclear energy industry. High-energy ion irradiations have been carried out at different accelerator facilities to induce the irradiation damage in these two materials; then, the mechanisms of damage have been characterized using principally Ion Beam Analysis (IBA) techniques.

One of the challenges of this thesis was to develop the ion beam induced luminescence or ionoluminescence (which is not a widely known IBA technique) and to apply it to the study of the mechanisms of irradiation damage in materials, proving the power of this technique. For this purpose, the ionoluminescence of three different types of silica (containing different amounts of OH groups) has been studied in detail and used to describe the creation and evolution of point defects under irradiation. In the case of MgO, the damage produced under 1.2 MeV Au$^+$ irradiation has been characterized using Rutherford backscattering spectrometry in channeling configuration and X-ray diffraction. Finally, the ionoluminescence of MgO under different irradiation conditions has also been studied.

The results obtained in this thesis help to understand the irradiation damage processes in materials, which is essential for the development of new nuclear energy sources.

Scientific Output

The work presented in this thesis has led to the following scientific output:

Publications

1. **D. Bachiller-Perea**, A. Muñoz-Martín, P. Corvisiero, D. Jiménez-Rey, V. Joco, A. Maira, A. Nakbi, A. Rodríguez, J. Narros, A. Zucchiatti. New Energy Calibration of the CMAM 5 MV Tandem Accelerator. *Energy Procedia*, 41:57–63, 2013. DOI: https://doi.org/10.1016/j.egypro.2013.09.007
2. **D. Bachiller-Perea**, D. Jiménez-Re, A. Muñoz-Martín, F. Agulló-López. Ion beam induced luminescence in amorphous silica: Role of the silanol group content and the ion stopping power. *Journal of Non-Crystalline Solids*, 428:36–41, 2015. DOI: https://doi.org/10.1016/j.jnoncrysol.2015.08.002
3. **D. Bachiller-Perea**, A. Debelle, L. Thomé, J.P. Crocombette. Study of the initial stages of defect generation in ion-irradiated MgO at elevated temperatures using high-resolution X-ray diffraction. *Journal of Materials Science*, 51:1456–1462, 2016. DOI: https://doi.org/10.1007/s10853-015-9465-3
4. **D. Bachiller-Perea**, D. Jiménez-Rey, A. Muñoz-Martín, F. Agulló-López. Exciton mechanisms and modeling of the ionoluminescence in silica. *Journal of Physics D: Applied Physics*, 49:085501, 2016. DOI: https://doi.org/10.1088/0022-3727/49/8/085501
5. **D. Bachiller-Perea**, A. Debelle, L. Thomé, M. Behar. Damage accumulation in MgO irradiated with MeV Au ions at elevated temperatures. *Journal of Nuclear Materials*, 478:268–274, 2016. DOI: https://doi.org/10.1016/j.jnucmat.2016.06.003
6. **D. Bachiller-Perea**, P. Corvisiero, D. Jiménez Rey, V. Joco, A. Maira Vidal, A. Muñoz Martin, A. Zucchiatti. Measurement of gamma-ray production X-sections in Li and F induced by protons from 810 keV to 3700 keV. *Nuclear*

Instruments and Methods in Physics Research Section B: Beam Interactions with Materials and Atoms, 406(A):161–166, 2017. DOI: https://doi.org/10.1016/j.nimb.2017.02.017

7. D. Jiménez-Rey, M. Benedicto, A. Muñoz-Martín, **D. Bachiller-Perea**, J. Olivares, A. Climent-Font, B. Gómez-Ferrer, A. Rodríguez, J. Narros, A. Maira, J. Àlvarez, A. Nakbi, A. Zucchiatti, F. de Aragón, J.M. García, R. Vila. First tests of the ion irradiation and implantation beamline at the CMAM. *Nuclear Instruments and Methods in Physics Research Section B: Beam Interactions with Materials and Atoms*, 331:196–203, 2014. DOI: https://doi.org/10.1016/j.nimb.2014.01.030

8. L. Beck, Y. Serruys, S. Miro, P. Trocellier, E. Bordas, F. Leprêtre, D. Brimbal, T. Loussarn, H. Martin, S. Vaubaillon, S. Pellegrino, **D. Bachiller-Perea**. Ion irradiation and radiation effect characterization at the JANNUS-Saclay triple beam facility. *Journal of Materials Research (Cambridge Journals)*, 2015. DOI: https://doi.org/10.1557/jmr.2014.414

9. A. Debelle, J.-P. Crocombette, A. Boulle, A. Chartier, T. Jourdan, S. Pellegrino, **D. Bachiller-Perea**, D. Carpentier, J. Channagiri, T.-H. Nguyen, F. Garrido, L. Thomé. Lattice strain in irradiated materials unveils a prevalent defect evolution mechanism. *Physical Review Materials*, 2:013604, 2018. DOI: https://doi.org/10.1103/PhysRevMaterials.2.013604

10. A. Debelle, J.-P. Crocombette, A. Boulle, E. Martinez, B. P. Uberuaga, **D. Bachiller-Perea**, Y. Haddad, F. Garrido, L. Thomé, and M. Béhar. How relative defect migration energies drive contrasting temperature-dependent microstructural evolution in irradiated ceramics. Physical Review Materials 2:083605, 2018. DOI: https://doi.org/10.1103/PhysRevMaterials.2.083605

Contributions to Conferences

1. **D. Bachiller-Perea**, D. Jiménez-Rey, A. Muñoz-Martín. Poster presentation: Radiation induced diffusion of light ions in insulators. 18th International Conference on Ion Beam Modification of Materials (IBMM 2012). Qingdao, China, 2–7 September 2012.

2. D. Jiménez-Rey, A. Muñoz-Martín, **D. Bachiller-Perea**, J. Olivares, A. Climent-Font, M.L. Crespillo, B. Gómez-Ferrer, A. Rodríguez, J. Narros, A. Maira, J. Álvarez, N. Nakbi, F. de Aragón, J.M. García, R. Vila, A. Zucchiatti. Poster presentation: The new ion irradiation and implantation beam line at CMAM. International workshop on the Modification and Analysis of Materials for Future Energy Sources (ENERGY-2012). Madrid, Spain, 17–20 September 2012.

3. **D. Bachiller-Perea**, D. Jiménez-Rey, V. Joco, A. Muñoz-Martín. Poster presentation: Radiation induced diffusion of light ions in insulators. International workshop on the Modification and Analysis of Materials for Future Energy Sources (ENERGY-2012). Madrid, Spain, 17–20 September 2012.

4. **D. Bachiller-Perea**, A. Muñoz-Martín, P. Corvisiero, D. Jiménez-Rey, V. Joco, A. Maira, et al. Contribution: New Energy Calibration of the CMAM 5MV Tandem Accelerator. International workshop on the Modification and Analysis of Materials for Future Energy Sources (ENERGY-2012). Madrid, Spain, 17–20 September 2012.

5. D. Jiménez-Rey, A. Muñoz Martín, **D. Bachiller-Perea**, J. Olivares, A. Climent Font, M. Crespillo Almenara, B. Gómez-Ferrer, A. Rodríguez, J. Narros, A. Maira, J. Álvarez, A. Nakbi, F. de Aragón, J. M. García, R. Vila, A. Zucchiatti. Poster presentation: The new ion irradiation and implantation beam line at CMAM. 21st International Conference on Ion Beam Analysis (IBA 2013). Seattle, WA, USA, June 2013.

6. **D. Bachiller-Perea**, A. Muñoz-Martín, F. Agulló-López, D. Jiménez-Rey. Poster presentation: Ion beam induced luminescence in fused silica: role of the stopping power. 19th International Conference on Ion Beam Modification of Materials (IBMM 2014). Leuven, Belgium, 14–19 September 2014.

7. **D. Bachiller-Perea**, D. Jiménez-Rey, A. Debelle, F. Agulló-López, A. Muñoz-Martín. Oral presentation: Ionoluminescence de la silice amorphe sous irradiations séquentielles. Ion Beam Analysis Francophone, 5me Rencontre "Analyse par faisceaux d'ions rapides" (IBAF 2014). Obernai, France, 7–10 October 2014. Special Mention of the Jury.

8. **D. Bachiller-Perea**, D. Jiménez-Rey, A. Debelle, A. Muñoz-Martín, F. Agulló-López. Poster presentation: Le rôle du pouvoir d'arrêt et du contenu en OH dans l'ionoluminescence de la silice amorphe. Ion Beam Analysis Francophone, 5me Rencontre "Analyse par faisceaux d'ions rapides" (IBAF 2014). Obernai, France, 7–10 October 2014.

9. **D. Bachiller-Perea**, A. Muñoz-Martín, D. Jiménez-Rey, A. Debelle, F. Agulló-López. Oral presentation: Ionoluminescence as a sensor of the defects creation and damage kinetics: application to fused silica. 22nd International Conference on Ion Beam Analysis (IBA 2015). Opatija, Croatia, 14–19 June 2015.

10. **D. Bachiller-Perea**, L. Thomé, A. Debelle. Poster presentation: Response of MgO to ion irradiation in the nuclear energy-loss regime and effect of irradiation temperature. 22nd International Conference on Ion Beam Analysis, (IBA 2015). Opatija, Croatia, 14–19 June 2015. Best Poster Award.

11. **D. Bachiller-Perea**, P. Corvisiero, D. Jiménez Rey, V. Joco, A. Maira Vidal, A. Muñoz Martin, A. Zucchiatti. Poster presentation: Measurement of gamma-ray production X-sections in Li and F induced by protons from 810 keV to 3700 keV. 12th European Conference on Accelerators in Applied Research and Technology (ECAART12). Jyväskylä, Finland, 3–8 July 2016.

Contents

Acronyms

Ciemat	Centro de Investigaciones Energéticas, Medioambientales y Tecnológicas
CL	CathodoLuminescence
CMAM	Centro de Micro-Análisis de Materiales
CNA	Centro Nacional de Aceleradores
CNRS	Centre National de la Recherche Scientifique
CSNSM	Centre de Sciences Nucléaires et de Sciences de la Matière
DEMO	DEMOnstration Power Plant
DFT	Density Functional Theory
DONES	Demo Oriented NEutron Source
EPR	Electron Paramagnetic Resonance
ERDA	Elastic Recoil Detection Analysis
ESR	Electron Spin Resonance
EVEDA	Engineering Validation and Engineering Design Activities
FWHM	Full Width at Half Maximum
HRXRD	High-Resolution X-Ray Diffraction
HWHM	Half Width at Half Maximum
IBA	Ion Beam Analysis
IBD	Ion Beam Deposition
IBIL	Ion Beam Induced Luminescence or ionoluminescence
IBML	Ion Beam Materials Laboratory
IBMM	Ion Beam Modification of Materials
ICP/AES	Inductively Coupled Plasma Atomic Emission Spectroscopy
IFMIF	International Fusion Materials Irradiation Facility
IL	Ion beam induced luminescence or IonoLuminescence
ITER	International Thermonuclear Experimental Reactor
JANNuS	Joint Accelerators for Nano-science and Nuclear Simulation
JET	Joint European Torus
LEIB	Low-Energy Ion Bombardment
LNT	Liquid-Nitrogen Temperature

MD	Molecular Dynamics
NBOHC	Non-Bridging Oxygen Hole Center
NRA	Nuclear Reaction Analysis
ODC	Oxygen Deficient Center
ODMR	Optically Detected Magnetic Resonance
PIGE	Particle Induced γ-ray Emission
PIXE	Particle Induced X-ray Emission
PL	PhotoLuminescence
pnA	Particle nanoAmpere
POL	Peroxy Linkage
POR	Peroxy Radical
ppm	Parts per million
RBS	Rutherford Backscattering Spectrometry
RBS/C	Rutherford Backscattering Spectrometry in Channeling configuration
RL	RadioLuminescence
RT	Room Temperature
SHI	Swift Heavy Ions
SRIM	The Stopping and Range of Ions in Matter
STE	Self-Trapped Exciton
STH	Self-Trapped Hole
TEM	Transmission Electron Microscopy
TL	ThermoLuminescence
ToF	Time of Flight
TRIM	The Transport of Ions in Matter
UAM	Universidad Autónoma de Madrid
UHV	Ultra High Vacuum
XRD	X-Ray Diffraction

Chapter 1
Introduction

1.1 Motivation of the Thesis

One of the most important issues that Physicists and Engineers face today is the development of a clean, sustainable, and efficient energy source that can satisfy the present and future needs of the society producing the minimum impact on the environment. For such a purpose, a huge international research effort is being devoted to the study of new systems of energy production, in particular in the field of nuclear energy. On the one hand, new fission reactors (Generation IV reactors, Gen IV) based on innovative concepts are being developed. On the other hand, today one of the biggest challenges from the technological point of view is the development of nuclear fusion reactors.

In fission reactors, nuclear chain reactions occur due to the interaction of neutrons with fissile isotopes. The nuclear fuel mostly used worldwide is uranium dioxide (UO_2) containing an enriched level of ^{235}U (3%, the 97 other % being ^{238}U) which is the only natural fissile element. Mixed uranium and plutonium (^{239}Pu) oxides, named MOX, are also used in some countries, particularly in France. The typical nuclear reactions produced with these isotopes are:

$$^{235}U + n \rightarrow Fission\ Fragments + 2.4\,n \tag{1.1}$$

$$^{239}Pu + n \rightarrow Fission\ Fragments + 2.9\,n \tag{1.2}$$

where the average values of the produced neutrons (n) in the different possible fission reactions have been considered. The average energy produced in these reactions is 192.9 MeV in the case of ^{235}U and 198.5 MeV for ^{239}Pu. The fission fragments produced in these reactions have an atomic mass in the ranges of 80–110 and 125–155 (e.g., ^{89}Kr, ^{144}Ba,...).

The Generation IV reactors will have major advantages compared to Generation II and III fission reactors. Among the six retained concepts, those involving fast neutrons (with either gas or sodium coolant) are the most promising ones. Indeed,

© Springer Nature Switzerland AG 2018
D. Bachiller Perea, *Ion-Irradiation-Induced Damage in Nuclear Materials*,
Springer Theses, https://doi.org/10.1007/978-3-030-00407-1_1

they should be able to use almost all plutonium isotopes as a fuel, to regenerate their fuel by producing fissile ^{239}Pu from fertile ^{238}U (with the use of fertile blankets in the core of the reactor), to produce less nuclear wastes with a lower radiotoxicity and to incinerate minor actinides. They will also be advantageous in terms of efficiency, sustainability, and safety [1].

In the case of fusion energy, the opposite type of nuclear reaction as compared to fission, is used to produce energy: light nuclei react forming heavier elements and emitting a large amount of energy. The most favorable fusion reaction in terms of cross-section and energy production is the following:

$$D + T \rightarrow n \ (14.03 \ MeV) + \alpha \ (3.56 \ MeV) \tag{1.3}$$

where D and T stand for deuterium and tritium, respectively (^{2}H and ^{3}H), and α stands for an alpha particle (^{4}He^{2+}). One gramme of this fuel can produce as much energy as 10 tons of petroleum or 1 kg of uranium. Fusion has other advantages such as being a quasi-unlimited source of energy (the Li resources are the limiting factor but they are estimated to be 1500 years), having a high efficiency, and producing a small quantity of nuclear wastes, mainly weakly activated materials due to neutron bombardment. However, to produce the energy in an efficient way the fuel needs to be heated at very high temperatures (10^8 K) where it reaches the plasma state. Then, the plasma has to be confined during a given time (that depends on the confinement method) to maintain the required conditions for the fusion reaction to start. At a certain point, the fusion reactions produce enough energy to maintain the temperature of the fuel, this is called the plasma ignition. Reaching ignition is an essential condition for the operation of a fusion reactor.

There are three possible ways of confining the plasma. The first one is the gravitational confinement; this is how the plasma is confined in stars, but it is not possible on Earth. The second method is the magnetic confinement in which powerful magnetic fields are applied to confine the plasma. The third method is inertial confinement, which consists in bombarding a target containing the nuclear fuel with multiple ($\sim$200) lasers simultaneously (or with electric discharges an in the Z machine [2]), the target being thus compressed and heated, producing the fusion of the fuel nuclei.

The magnetic confinement is the most developed confinement method today, and the most likely to be used in the future fusion reactors. The two main devices to produce magnetic confinement are *tokamaks* and *stellerators*. In *tokamaks* (from the Russian, "toroidal chamber with magnetic coils") two powerful magnetic fields (toroidal and poloidal) are combined to reach the optimal configuration to confine the plasma, the poloidal field is generated by a high electric current circulating in the plasma, this current can provoke instabilities in the plasma (disruptions). In the case of *stellerators* the magnetic field is only produced by coils, no current circulates in the plasma, avoiding the disruptions; however, the configuration of the coils is much more complicated in this case.

The viability of producing fusion energy was demonstrated with the tokamak reactor JET (Joint European Torus, Oxford, United Kingdom) in 1991 [3], the applied power was 22.8 MW and it produced 16 MW, which means a Q-factor (gain factor, ratio between the power produced by the reactor and the power applied to maintain the plasma) of ~0.7. The condition of $Q = 1$ is referred to as *breakeven*, a Q-factor of ~5 is required to reach ignition, and a commercial reactor will need a Q-factor of ~20.

Currently, the *tokamak* reactor ITER (International Thermonuclear Experimental Reactor) is being built in Cadarache, France [4]. ITER is an experimental reactor aiming at testing various technologies to improve, e.g., the magnetic coils efficiency, the plasma stability or the material resistance. It is most of all designed to sustain plasma discharges as long as 500 s for a produced fusion power of 400 MW with a Q factor of 10.

After the development of ITER, another nuclear fusion power station is intended to be built before the construction of the industrial prototypes: DEMO (DEMOnstration Power Station). The materials used in this facility will be exposed to a maximum dose of 20 dpa (displacements per atom) in a first phase of operation, and to 50 dpa in a second phase [5].

The first industrial prototype of a nuclear fusion power station is intended to be PROTO, which will be implemented not before 2050.

Concurrently with these advancements in fusion reactors, other projects regarding the neutron irradiation of materials are being developed. This is the case of the IFMIF (International Fusion Materials Irradiation Facility), IFMIF/EVEDA (Engineering Validation and Engineering Design Activities), and DONES (Demo Oriented NEutron Source) projects. These projects consists in the development of accelerator-based neutron sources that will produce a large neutron flux with a spectrum similar to that expected at the first wall of a fusion reactor. The main goal of these facilities is to test radiation-resistant and low-activation materials that could withstand the high neutron fluxes in a fusion reactor (see [5] and references therein).

The study of the irradiation damage in materials is one of the main important issues for the development of the future Generation IV and fusion reactors [6]. This study can be approached in two different ways; one is from a technological and applied point of view, the second is from a fundamental point of view to understand the mechanisms of damage that take place in materials under irradiation. Since the neutron irradiation facilities with the same irradiation conditions than fusion reactors (such as IFMIF) have not been built yet, a possible way to emulate the neutron damage in materials is by performing ion irradiations. In particular, it has been proposed that the neutron damage can be emulated by bombarding materials with a triple ion beam including light ions (H+He) and heavy ions [7, 8]. Only a few experimental facilities in the world can perform triple ion beam irradiations, one of them being the JANNuS facility in Saclay, France [9].

The use of tools such as ion accelerators is essential to understand the mechanisms of damage that take place in materials under irradiation. Ion accelerators produce highly-energetic ion-beams (in the range of MeV/nucleon) that can be used to simulate the effects of the irradiation in nuclear reactors. Moreover, with the appropriate experimental techniques, it is possible to characterize the materials that could be used in the future nuclear reactors and to understand the mechanisms of transformation (most generally damage) by irradiation in these materials at both microscopic and macroscopic scales.

In this thesis two materials have been studied: amorphous silica (a-SiO$_2$) and magnesium oxide (MgO). Both materials are oxides with applications in the nuclear energy industry, therefore, they will be subjected to high levels of radiation, making necessary the study of their behavior under irradiation. Although both a-SiO$_2$ and MgO are insulating oxide materials, their structure is very different: a-SiO$_2$ is amorphous and has a pronounced covalent character, while MgO is crystalline and has a very ionic character. Due to these differences, different techniques have to be used to study the damage produced in each material by irradiation.

Silicon dioxide (SiO$_2$) can be found in the nature either in crystalline (quartz) or in amorphous form (fused silica, a-SiO$_2$). Both materials are used for several technological applications. In particular, amorphous silica is vastly used for optical and electrical devices. In some of these applications this material is exposed to hostile environments (high levels of radiation, extreme temperatures,...) as, for instance, when it is used in devices in space platforms, in laser applications, and in nuclear fission and fusion facilities [10–12]. It is foreseen that fused silica will be a functional material in ITER and DEMO fusion reactors for diagnosis, remote handling and optical devices. The extreme conditions of radiation in such fusion reactors are expected to produce a high radiation-induced damage in structural and functional materials. It is therefore essential to know the mechanisms of damage formation in fused silica to understand its modifications at the molecular level.

On the other hand, MgO is a very good contender as a neutron reflector in fast neutron reactors [13], it is envisaged as a matrix of the ceramic–ceramic composite nuclear fuel for minor actinide transmutation [14], and it could be also used as an electrical insulator for diagnostic components in the ITER fusion reactor [15], making it a potential candidate for applications in the nuclear energy field. The properties of MgO that make this material suitable for such nuclear applications are a low neutron capture cross section, a high thermal conductivity and a good radiation resistance. Under such operating conditions, MgO will definitely be subjected to various types of irradiation at high temperatures, typically in the range of 500–1200 K and that makes mandatory the good knowledge of the MgO behavior under ion irradiation at elevated temperatures. It is here noteworthy that MgO responds to ion irradiation in a non-trivial way due to the possible formation of charged defects on the two sublattices.

In this work, high-energy ion irradiations have been carried out at different accelerator facilities to induce the irradiation damage in these two materials; then, the mechanisms of damage have been characterized using principally Ion Beam Analysis (IBA) techniques. One of the main techniques used in this thesis is the Ion Beam Induced Luminescence or ionoluminescence (IBIL, IL), which is not a widely known IBA technique. One of the challenges of this thesis was to develop the IL technique under different irradiation conditions (different ions and temperatures, at different research centers,...) and to apply it to the study of the mechanisms of irradiation damage in materials, proving the power of this technique.

1.2 Current State of Knowledge and Issues Addressed in this Work

Irradiation-induced effects in materials include a wide range of changes due to the variety of ion-solid interactions (see review books [16, 17]); changes extend from point-defect creation [18] to complete phase transformation [19], including amorphization [20], but also comprise significant microstructural modifications such as polygonization or cracking [21, 22]. The irradiation effects depend on many parameters; firstly, they differ depending on the irradiation conditions, and principally on the type of slowing down of the ions [16, 17], but also on the nature of the ions (e.g., soluble or insoluble), as well as on the flux, fluence, and temperature (see e.g., [23]). Secondly, they depend on the intrinsic material properties, including the thermal and mechanical properties that are important in the energy dissipation process, and on the resistance of the material to defect incorporation. Indeed, the final state of an irradiated material depends heavily on the early irradiation stages, i.e., on type, density, and mobility of the primary defects. The characteristics of these defects are closely related to the electronic structure of the material, meaning that the response to irradiation of the material, metal or insulator, ionic or covalent bonding, is different [16, 17].

Since the field of irradiation damage in materials is very wide, this section will only focus on the works carried out on the two materials studied in this thesis: a-SiO$_2$ and MgO. Although there are already very important studies in this field, a lot of crucial issues remain to be solved, and the experimental studies presented in this thesis will help to understand the behavior of materials under irradiation.

1.2.1 Amorphous Silica

A lot of research activity has been devoted [12, 24–35] to understanding the atomic and electronic structures of the defects occurring in the crystalline and amorphous phases of SiO$_2$ and the effects of different types of irradiation [36–42]. So far, the effects of ion-beam irradiation have been mostly studied for light ions (H, He) [27, 31, 33, 43–48]. More recently, the focus of the research has shifted to the damage produced by high-energy heavy ions [35, 49–56] (so called swift heavy ions, SHI), where the electronic excitation, i.e., related to the electronic stopping power (see Sect. 3.1.1), reaches much higher rates than for light ions. This latter case represents a difficult scientific challenge and presents specific features that are not yet sufficiently understood [56–61]. It is increasingly realized that the processes induced during irradiation with SHI are very different from those operating under irradiation with lighter ions and should be associated with the relaxation of the excited electronic system.

Luminescence is a very sensitive technique for identifying and investigating optically active point defects (color centers) in dielectric materials [62, 63]. In particular, for SiO_2 the correlation between optical absorption and photoluminescence (PL) has allowed the identification of a number of relevant color centers [32, 64, 65], although much additional work is still required to understand in detail the behavior of this important material when it is irradiated. On the other hand, luminescence during irradiation is a useful tool for investigating the generation of point defects through irradiation and revealing the operative mechanisms. An increasing number of papers has recently been devoted to identifying the defects produced in SiO_2 by irradiation with high-energy ion beams and understanding their correlation with the associated luminescence processes (ionoluminescence, IL) [35, 54, 55]. The light emission mechanisms are, indeed, of an electronic nature and so they may help to clarify the processes induced by the electronic excitation and highlight the differences with the elastic collision processes. One main intrinsic advantage of the IL technique over other spectroscopic methods (such as optical absorption, Raman, and electronic paramagnetic resonance) is that spectra are obtained in situ, not requiring the interruption of the irradiation and so avoiding the influence of possible recovery effects [66, 67]. Although recent progresses have been reported on the features of IL emission in SiO_2 (crystalline and amorphous) and its relation to radiation-induced processes, a coherent understanding of the IL mechanisms is not yet available. Since a better knowledge of the structure and spectroscopy of the created defects has recently been achieved (see Sect. 2.1.2), a deeper physical discussion of the IL processes has now become possible.

One of the purposes of this thesis is to report new experimental results comparing the IL kinetic behavior obtained under light- and heavy-ion irradiation in order to clarify the physical processes operating in each case. This comparison is relevant since, so far, most experiments have been performed at room temperature (RT) using light ions. In all cases two main bands are observed, one appearing at 1.9 eV and generally associated with non-bridging oxygen hole (NBOH) centers, and another one in the blue spectral region at around 2.7 eV. The origin of this later band is still a matter of conflict. It has been sometimes attributed to intrinsic recombination of self-trapped excitons (STEs) [68, 69], in accordance with some low temperature irradiation experiments [25, 70, 71] and several theoretical analyses [53, 68, 69, 72]. However, other authors, performing RT irradiations with light ions (mostly hydrogen), assigned the emission to electron-hole recombination at Oxygen-Deficient Centers (ODCs) [27, 31, 33].

It is noteworthy that the role of the STE migration through the material network has been so far weakly considered, except by a work by Costantini et al. [35], performed on natural quartz at liquid-nitrogen temperature (LNT).

The evolution of the silica network structure with irradiation fluence has been investigated by infrared (IR) [53] and Raman [73] spectroscopies. In fact, the IL kinetic behavior observed in our experiments has been well correlated with the radiation-induced evolution of the first-order vibrational peak [53] appearing at the mode frequency ω_4. These optical data account for the decrease in the radius of the rings and the associated compaction of the material as a main macroscopic result induced by ion-beam irradiation. As a consequence of our work we can offer a real-time tool (IL) to investigate the synergy between the structural damage (compaction) and the coloring mechanisms in silica.

1.2.2 Magnesium Oxide

As explained in Sect. 1.1, in operating conditions, MgO will necessarily be exposed to ion irradiation at high temperatures, in particular in the 500–1200 K range. However, studies on MgO dealing with the effect of the irradiation temperature are scarce in this temperature range. One can nevertheless cite a few significant experimental works conducted with different irradiation sources. For instance, MgO has been studied under highly-energetic electron irradiation from room temperature (RT) to 1273 K [74–76], under fast neutron irradiation at 923 K [77], and under different ion irradiation conditions from below RT to 1373 K [78–81]. Several key results were hence obtained. Irrespective of the irradiation conditions, point defects are formed in MgO at low irradiation dose. Sambeek et al. [78] established a quantitative correlation between lattice parameter change and point-defect concentration in the early stages of irradiation under various conditions. With increasing dose, production of $1/2\langle110\rangle\{100\}$ interstitial dislocation loops is observed. The dislocation loop nucleation results from the clustering of point defects. The dislocation loop density was found to depend on the temperature and, more precisely, on the relative mobility of point defects. In fact, the nucleation mechanism below 873 K is controlled by interstitial motion in which one or two pairs of Mg- and O-interstitials serve as stable nuclei for interstitial loops; above 873 K, the growth kinetics of loops is explained in terms of the steady state behavior of high mobility interstitials and vacancies, leading to a sharp decrease in the loop density [76]. No void formation has been reported under neutron and ion irradiation. In a more recent study, Usov et al. [80] aimed at monitoring, using Rutherford backscattering spectroscopy in channeling configuration (RBS/C), the disorder level in MgO irradiated with 100 keV Ar at temperatures from 123 to 1373 K. They pointed out a partial damage recovery (at least in {100}-oriented crystals) over the whole investigated temperature range.

Defect energetics (and hence defect migration properties) has been the focus of several, relatively recent computational works. It was demonstrated that both Mg and O vacancies are immobile at RT, with migration energies reaching a few eV [82–84]. On the contrary, for O and Mg mono-interstitials low migration barriers were found, with values ranging from 0.3 to 0.7 eV, depending on the computational method and on the charge state of the defects; for the O^{2-} mono-interstitial, calculations even resulted in an extremely low value of 0.06 eV [84]. Uberuaga et al. [82] showed the formation of mono-interstitials and mono-vacancies for the most part after a collision cascade, while increasing recoil energy resulted in the formation of slightly bigger defects. They also forecast in agreement to experimental observations on nucleation of dislocation loops that interstitials tend to agglomerate when their density increases and larger clusters become more and more stable.

In conclusion, most of the experimental works on MgO under irradiation aimed at investigating irradiation stages where extended defects are formed. The defect mobility was put forward as a key parameter in explaining the obtained results; this assumption was backed up by computational studies that focused on the early formation of primary defects and on their clustering properties. Nevertheless, there is a lack of basic data regarding the damage accumulation process in MgO irradiated at elevated temperatures. Similarly, a limited number of experimental works have been focused on the early stages of irradiation effects in MgO. To address these questions, in this thesis, the RBS/C technique has been used to obtain the disorder depth profiles and the damage accumulation in single-crystalline MgO samples irradiated with 1.2 MeV Au^+ ions at 573, 773, and 1073 K and at different fluences (more precisely, until saturation of the disorder). In addition, the initial stages of defect generation in MgO under these irradiation conditions have been studied. High-resolution X-ray diffraction has been used to measure the irradiation-induced elastic strain. Point-defect relaxation volumes have been computed using density functional theory calculations and the defect concentration has been calculated.

The ionoluminescence of MgO has also been studied in this thesis. The information that can be found in the literature about the IL of MgO is very scarce since it has almost not been studied. Three of the very limited articles published about the IBIL of MgO are [85–87], and only one IL spectrum is shown in these three papers. However, the luminescence of MgO has been studied with other techniques such as thermoluminescence (TL) [88–91], cathodoluminescence (CL) [90, 92] or photoluminescence (PL) [93–97]. These studies have provided information about the main luminescence emissions of MgO and their possible origin; there are also some databases that compile this information [98, 99]. However, some of the reported data are sometimes controversial. In order to provide new, and tentatively conclusive data on the IL of MgO, measurements at 100 and 300 K under light- and heavy-ion irradiation have been performed in this thesis; the results obtained here provide insights about the origin of the main luminescence emissions in MgO.

1.3 Description of the Chapters

This thesis is divided into three parts. Part I describes the materials and methods used in this work, and is made up of Chaps. 2, 3, 4, and 5. Part II (Chaps. 6, 7, and 8) contains all the results obtained for the ionoluminescence in amorphous silica. The study of the ion-irradiation damage in MgO is presented in Part III, which consists of Chaps. 9 and 10.

Chapter 2 describes the properties of the two materials studied in this thesis: amorphous silica (a-SiO$_2$) and magnesia (magnesium oxyde, MgO). The characteristics of the samples used in our experiments (structure, impurities, preparation of the samples,...) are presented in this chapter. A special emphasis is made on the type of defects that can be found in both materials, since the main results of this thesis deal with the defects created in silica and MgO as a consequence of the ion irradiation.

Chapter 3 summarizes the main concepts of the ion-solid interactions, which are essential to understand the techniques used in this thesis and the results obtained. First, concepts such as nuclear and electronic stopping powers, ion range, or defect creation mechanisms are introduced here. Then, the main characteristics of the SRIM program, which is very often used in the field of ion irradiation of materials, are described. Finally, the main processes of ion beam modification of materials and the ion beam analysis techniques are presented.

The main accelerator facilities in which the experiments of the thesis have been carried out are presented in Chap. 4. A brief introduction to the principal elements of this type of facilities is done. The three experimental facilities described here are the Centro de Micro-Análisis de Materiales (CMAM, Madrid, Spain), the Centre de Sciences Nuclèaires et de Sciences de la Matière (CSNSM, Orsay, France), and the Ion Beam Materials Laboratory (IBML, Knoxville, United States).

Chapter 5 describes the main experimental techniques used for this work. Although other techniques have been used to obtain complementary information of the samples (such as optical absorption or particle-induced X-ray emission, PIXE), the three principal experimental techniques that have been used are: ion beam induced luminescence, Rutherford backscattering spectrometry, and X-ray diffraction.

In Chap. 6 the general features of the ionoluminescence of amorphous silica are presented. The origin of the main IL bands of a-SiO_2 is explained; IL experiments at low temperature help to understand the origin of these emissions. The nuclear contribution to the IL signal is also studied here.

Chapter 7 provides a novel set of data of the ionoluminescence of silica at room temperature for different ions and energies covering a large range of stopping powers. The results are compared for three types of silica containing different amounts of OH groups, which is a typical dopant used to modify the optical properties of a-SiO_2. The dependence of the IL emissions on the OH content of the samples and on the stopping power of the incident ions is studied.

A physical model is proposed in Chap. 8 to explain the evolution of the IL emissions of silica with the irradiation fluence. From this model, a mathematical formulation is derived and used to fit the experimental IL data.

The study of the ion irradiation damage produced in MgO by 1.2 MeV Au^+ irradiation at high temperatures is presented in Chap. 9. The damage produced by the irradiation at three temperatures (573, 773, and 1073 K) is characterized by Rutherford backscattering spectrometry in channeling configuration (RBS/C), and X-ray diffraction (XRD). The defect generation and the damage accumulation processes are studied.

Chapter 10 provides new results on the ionoluminescence of MgO at low temperature (100 K) and at room temperature (300 K) with light ions (H) and with heavy ions (Br). The emissions observed in the four cases are reported, and a preliminary interpretation of the IL emissions is done (although more data are required to better determine their origin since the ionoluminescence of MgO has practically not been studied before).

Finally, Chap. 11 summarizes the main conclusions of the thesis and suggests the future research lines that could be followed to continue this work.

References

1. Technology roadmap update for generation IV nuclear energy systems, OECD nuclear energy agency for the generation IV international forum, 2014. https://www.gen-4.org/gif/upload/docs/application/pdf/2014-03/gif-tru2014.pdf
2. Z Machine, Sandia national laboratories, Alburquerque, New Mexico, USA. http://www.sandia.gov/z-machine/
3. JET, Joint European torus. www.euro-fusion.org/jet/
4. ITER, International thermonuclear experimental reactor. www.iter.org
5. F. Mota, Á. Ibarra, Á. García, J. Molla, Sensitivity of IFMIF-DONES irradiation characteristics to different design parameters. Nucl. Fusion **55**(12), 123024 (2015)
6. L.K. Mansur, A.F. Rowcliffe, R.K. Nanstad, S.J. Zinkle, W.R. Corwin, R.E. Stoller, Materials needs for fusion, generation IV fission reactors and spallation neutron sources - similarities and differences. J. Nucl. Mater. **329–333**, Part A:166–172 (2004). Proceedings of the 11th International Conference on Fusion Reactor Materials (ICFRM-11)
7. D. Jiménez-Rey, F. Mota, R. Vila, A. Ibarra, Christophe J. Ortiz, J.L. Martínez-Albertos, R. Román, M. González, I. García-Cortes, J.M. Perlado, Simulation for evaluation of the multi-ion-irradiation Laboratory of TechnoFusión facility and its relevance for fusion applications. J. Nucl. Mater. **417**(1–3):1352 – 1355 (2011). Proceedings of ICFRM-14
8. TechnoFusión, National Centre for Fusion Technologies, Scientific-Technical Report, 2009. http://www.technofusion.org/documents/TF_Report_ENG.pdf
9. JANNuS - Joint accelerators for nano-science and nuclear simulation, France. http://jannus.in2p3.fr/spip.php
10. A. Ibarra, E.R. Hodgson, The ITER project: the role of insulators. Nucl. Instrum. Methods Phys. Res. Sect. B: Beam Interact. Mater. Atoms **218**, 29–35 (2004)
11. A. Moroño, E.R. Hodgson, Radiation induced optical absorption and radioluminescence in electron irradiated SiO_2. J. Nucl. Mater. **258–263**(2), 1889–1892 (1998)
12. J.F. Latkowski, A. Kubota, M.J. Caturla, S.N. Dixit, J.A. Speth, S.A. Payne, Fused silica final optics for inertial fusion energy: radiation studies and system-level analysis. Fusion Sci. Technol. **43**(4), 540–558 (2003)
13. R.R. Macdonald, M.J. Driscoll, Magnesium oxide: an improved reflector for blanket-free fast reactors. Trans. Am. Nucl. Soc. **102**, 488–489 (2010)
14. S. Somiya, *Handbook of advance ceramics: materials, applications, processing and properties* (Academic, Elsevier, Oxford, 2013)
15. T. Shikama, T. Nishitani, T. Kakuta, S. Yamamoto, S. Kasai, M. Narui, E. Hodgson, R. Reichle, B. Brichard, A. Krassilinikov, R. Snider, G. Vayakis, A. Costley, S. Nagata, B. Tsuchiya, K. Toh, Irradiation test of diagnostic components for ITER application in the Japan materials testing reactor. Nucl. Fusion **43**(7), 517–521 (2003)
16. M. Nastasi, J.W. Mayer, J.K. Hirvonen, in *Ion-solid Interactions Fundamentals and Applications*. Cambridge Solid State Science Series (Cambridge University Press, Cambridge, 1996)
17. K.E. Sickafus, E.A. Kotomin, B.P. Uberuaga (eds.), in *Radiation Effects in Solids*. Nato Sciences Series vol. 235 (Springer, Berlin, 2007)
18. J.P. Rivière, in *Application of Particle and Laser Beams in Materials Technology*, Radiation induced point defects and diffusion. Nato Science Series E, vol. 283 (Springer, Berlin, 1995), pp. 53–76
19. M. Lang, R. Devanathan, M. Toulemonde, C. Trautmann, Advances in understanding of swift heavy-ion tracks in complex ceramics. Curr. Opin. Solid State Mater. Sci. **19**(1), 39–48 (2015)
20. S. Moll, G. Sattonnay, L. Thomé, J. Jagielski, C. Decorse, P. Simon, I. Monnet, W.J. Weber, Irradiation damage in $Gd_2Ti_2O_7$ single crystals: ballistic versus ionization processes. Phys. Rev. B Condens. Matter Mater. Phys. **84**, 064115 (2011)
21. Hj Matzke, L.M. Wang, High-resolution transmission electron microscopy of ion irradiated uranium oxide. J. Nucl. Mater. **231**(1–2), 155–158 (1996)
22. G. Velisa, A. Debelle, L. Vincent, L. Thomé, A. Declémy, D. Pantelica, He implantation in cubic zirconia: deleterious effect of thermal annealing. J. Nucl. Mater. **402**(1), 87–92 (2010)

23. E. Wendler, B. Breeger, Ch. Schubert, W. Wesch, Comparative study of damage production in ion implanted III-V-compounds at temperatures from 20 to 420 K. Nucl. Instrum. Methods Phys. Res. Sect. B Beam Interact. Mater. Atoms **147**(1–4), 155–165 (1999)

24. K. Tanimura, T. Tanaka, N. Itoh, Creation of quasistable lattice defects by electronic excitation in SiO_2. Phys. Rev. Lett. **51**(5), 423–426 (1983)

25. C. Itoh, K. Tanimura, N. Itoh, Optical studies of self-trapped excitons in SiO_2. J. Phys. C Solid State Phys. **21**(26), 4693–4702 (1988)

26. F. Agulló-López, C.R.A. Catlow, P.D. Townsend, *Point Defects in Materials* (Academic, San Diego, 1988)

27. P.D. Townsend, P.J. Chandler, L. Zhang, *Optical Effects of Ion Implantation* (Cambridge University Press, Cambridge, 1994)

28. D.L. Griscom, Gamma and fission-reactor radiation effects on the visible-range transparency of aluminum-jacketed, all-silica optical fibers. J. Appl. Phys. **80**, 2142–2155 (1996)

29. C.D. Marshall, J.A. Speth, S.A. Payne, Induced optical absorption in gamma, neutron and ultraviolet irradiated fused quartz and silica. J. Non-Cryst. Solids **2112**(1), 59–73 (1997)

30. N. Itoh, A.M. Stoneham, *Materials Modification by Electronic Excitation* (Cambridge University Press, Cambridge, 2001)

31. S. Nagata, S. Yamamoto, K. Toh, B. Tsuchiya, N. Ohtsu, T. Shikama, H. Naramoto, Luminescence in SiO_2 induced by MeV energy proton irradiation. J. Nucl. Mater. **329–333**(B), 1507–1510 (2004). Proceedings of the 11th International Conference on Fusion Reactor Materials (ICFRM-11)

32. L. Skuja, M. Hirano, H. Hosono, K. Kajihara, Defects in oxide glasses. Phys. Status Solidi (c) **2**(1), 15–24 (2005)

33. S. Nagata, S. Yamamoto, A. Inouye, B. Tsuchiya, K. Toh, T. Shikama, Luminescence characteristics and defect formation in silica glasses under h and he ion irradiation. J. Nucl. Mater. **367–370**(B), 1009–1013 (2007)

34. L. Skuja, K. Kajihara, M. Hirano, H. Hosono, Oxygen-excess-related point defects in glassy/amorphous SiO_2 and related materials. Nucl. Instrum. Methods Phys. Res. Sect. B Beam Interact. Mater. Atoms **286**, 159–168 (2012)

35. J.M. Costantini, F. Brisard, G. Biotteau, E. Balanzat, B. Gervais, Self-trapped exciton luminescence induced in alpha quartz by swift heavy ion irradiations. J. Appl. Phys. **88**, 1339–1345 (2000)

36. D. Bravo, J.C. Lagomacini, M. León, P. Martín, A. Martín, F.J. López, A. Ibarra, Comparison of neutron and gamma irradiation effects on KU1 fused silica monitored by electron paramagnetic resonance. Fusion Eng. Des. **84**(26), 514–517 (2009)

37. P. Martín, M. León, A. Ibarra, E.R. Hodgson, Thermal stability of gamma irradiation induced defects for different fused silica. J. Nucl. Mater. **417**(1–3), 818–821 (2011). Proceedings of ICFRM-14

38. M. León, P. Martín, A. Ibarra, E.R. Hodgson, Gamma irradiation induced defects in different types of fused silica. J. Nucl. Mater. **386–388**, 1034–1037 (2009). Proceedings of the Thirteenth International Conference on Fusion Reactor Materials

39. M. León, L. Giacomazzi, S. Girard, N. Richard, P. Martín, L. Martín-Samos, A. Ibarra, A. Boukenter, Y. Ouerdane, Neutron Irradiation effects on the structural properties of KU1, KS-4V and I301 silica glasses. IEEE Trans. Nucl. Sci. **61**(4), 1522–1530 (2014)

40. AKh Islamov, U.S. Salikhbaev, E.M. Ibragimova, I. Nuritdinov, B.S. Fayzullaev, KYu. Vukolov, I. Orlovskiy, Efficiency of generation of optical centers in KS-4V and KU-1 quartz glasses at neutron and gamma irradiation. J. Nucl. Mater. **443**(1–3), 393–397 (2013)

41. A. Ibarra, A. Muñoz-Martín, P. Martín, A. Climent-Font, E.R. Hodgson, Radiation effects on the deuterium diffusion in SiO_2. J. Nucl. Mater. **367–370**(Part B), 1003–1008 (2007). Proceedings of the Twelfth International Conference on Fusion Reactor Materials (ICFRM-12)

42. J.C. Lagomacini, D. Bravo, P. Martín, A. Ibarra, A. Martín, F.J. López, EPR study of new defects in neutron irradiated KS-4V and KU1 fused silica, in *IOP Conference Series: Materials Science and Engineering*, vol. 15, issue 1 (2010), pp. 012052

43. G.E. King, A.A. Finch, R.A.J. Robinson, D.E. Hole, The problem of dating quartz 1: spectroscopic ionoluminescence of dose dependence. Radiat. Meas. **46**(1), 1–9 (2011)
44. A.A. Finch, J. Garcia-Guinea, D.E. Hole, P.D. Townsend, J.M. Hanchar, Ionoluminescence of zircon: rare earth emissions and radiation damage. J. Phys. D Appl. Phys. **37**(20), 2795–2803 (2004)
45. J. Demarche, D. Barba, G.G. Ross, G. Terwagne, Ionoluminescence induced by low-energy proton excitation of Si nanocrystals embedded in silica. Nucl. Instrum. Methods Phys. Res. Sect. B Beam Interact. Mater. Atoms **272**, 141–144 (2012)
46. O. Kalantaryan, S. Kononenko, V.P. Zhurenko, Ionoluminescence of silica bombarded by 420 keV molecular hydrogen ions. Funct. Mater. **20**(4), 462–465 (2013)
47. O. Kalantaryan, S. Kononenko, V. Zhurenko, N. Zheltopyatova, Fast ion induced luminescence of silica implanted by molecular hydrogen. Funct. Mater. **21**(1), 26–30 (2014)
48. S.I. Kononenko, O.V. Kalantaryan, V.I. Muratov, V.P. Zhurenko, Features of silica luminescence induced by molecular hydrogen ions. Nucl. Instrum. Methods Phys. Res. Sect. B: Beam Interac. Mater. Atoms **246**(2), 340–344 (2006)
49. J. Manzano-Santamaría, J. Olivares, A. Rivera, O. Peña-Rodríguez, F. Agulló-López, Kinetics of color center formation in silica irradiated with swift heavy ions: thresholding and formation efficiency. Appl. Phys. Lett. **101**, 154103 (2012)
50. P. Martín, D. Jiménez-Rey, R. Vila, F. Sanchez, R. Saavedra, Optical absorption defects created in SiO2 by Si, O and He ion irradiation. Fusion Eng. Des. **89**, 1679–1683 (2014)
51. J. Manzano-Santamaría, Daño por excitación electrónica en SiO₂ mediante irradiaciones con iones pesados de alta energía, Ph.D. thesis, Universidad Autónoma de Madrid, 2013
52. M. Ma, X. Chen, K. Yang, X. Yang, Y. Sun, Y. Jin, Z. Zhu, Color center formation in silica glass induced by high energy fe and xe ions. Nucl. Instrum. Methods Phys. Res. Sect. B Beam Interact. Mater. Atoms **268**(1), 67–72 (2010)
53. K. Awazu, S. Ishii, K. Shima, S. Roorda, J.L. Brebner, Structure of latent tracks created by swift heavy-ion bombardment of amorphous sio₂. Phys. Rev. B Condens. Matter Mater. Phys. **62**, 3689–3698 (2000)
54. D. Jiménez-Rey, O. Peña-Rodríguez, J. Manzano-Santamaría, J. Olivares, A. Muñoz-Martín, A. Rivera, F. Agulló-López, Ionoluminescence induced by swift heavy ions in silica and quartz: a comparative analysis. Nucl. Instrum. Methods Phys. Res. Sect. B Beam Interact. Mater. Atoms **286**, 282–286 (2012). Proceedings of the Sixteenth International Conference on Radiation Effects in Insulators (REI)
55. O. Peña-Rodríguez, D. Jiménez-Rey, J. Manzano-Santamaría, J. Olivares, A. Muñoz, A. Rivera, F. Agulló-López, Ionoluminescence as sensor of structural disorder in crystalline sio2: determination of amorphization threshold by swift heavy ions. Appl. Phys. Express **5**(1), 011101 (2012)
56. L. Thomé, A. Debelle, F. Garrido, S. Mylonas, B. Décamps, C. Bachelet, G. Sattonnay, S. Moll, S. Pellegrino, S. Miro, P. Trocellier, Y. Serruys, G. Velisa, C. Grygiel, I. Monnet, M. Toulemonde, P. Simon, J. Jagielski, I. Jozwik-Biala, L. Nowicki, M. Behar, W.J. Weber, Y. Zhang, M. Backman, K. Nordlund, F. Djurabekova, Radiation effects in nuclear materials: role of nuclear and electronic energy losses and their synergy. Nucl. Instrum. Methods Phys. Res. Sect. B Beam Interact. Mater. Atoms **307**, 43–48 (2013)
57. Z.G. Wang, Ch. Dufour, E. Paumier, M. Toulemonde, The se sensitivity of metals under swift-heavy-ion irradiation: a transient thermal process. J. Phys. Condens. Matter **6**, 6733–6750 (1994)
58. A. Meftah, F. Brisard, J.M. Costantini, E. Dooryhee, M. Hage-Ali, M. Hervieu, J.P. Stoquert, F. Studer, M. Toulemonde, Track formation in SiO₂ quartz and the thermal-spike mechanism. Phys. Rev. B Condens. Matter Mater. Phys. **49**, 12457–12463 (1994)
59. S. Klaumünzer, Ion tracks in quartz and vitreous silica. Nucl. Instrum. Methods Phys. Res. Sect. B Beam Interac. Mater. Atoms **225**(1–2), 136–153 (2004). The Evolution of Ion Tracks and Solids (TRACKS03)
60. P. Kluth, C.S. Schnohr, O.H. Pakarinen, F. Djurabekova, D.J. Sprouster, R. Giulian, M.C. Ridgway, A.P. Byrne, C. Trautmann, D.J. Cookson, K. Nordlund, M. Toulemonde, Fine structure in swift heavy ion tracks in amorphous sio₂. Phys. Rev. Lett. **101**, 175503 (2008)

61. N. Itoh, D.M. Duffy, S. Khakshouri, A.M. Stoneham, Making tracks: electronic excitation roles in forming swift heavy ion tracks. J. Phys. Condens. Matter **21**, 474205 (2009)
62. R.C. Ropp, *Luminescence and the Solid State* (Elsevier, Amsterdam, 2004)
63. R. Sahl, in *Crystalline Silicon-Properties and Uses*. Defect related luminescence in silicon dioxide network: a review (InTech, Rijeka, Croatia, 2011), pp. 135–172
64. L. Skuja, Optically active oxygen-deficiency-related centers in amorphous silicon dioxide. J. Non-Cryst. Solids **239**(1–3), 16–48 (1998)
65. A.N. Trukhin, Luminescence of localized states in silicon dioxide glass. a short review. J. Non-Cryst. Solids **357**(8–9), 1931–1940 (2011). SiO2, Advanced Dielectrics and Related Devices
66. R.A.B. Devine, J. Arndt, Correlation defect creation and dose-dependent radiation sensitivity in amorphous SiO2. Phys. Rev. B Condens. Matter Mater. Phys. **39**, 5132–5138 (1989)
67. H. Hosono, K. Kajihara, T. Suzuki, Y. Ikuta, L. Skuja, M. Hirano, Vacuum ultraviolet optical absorption band of non-bridging oxygen hole centers in SiO2 glass. Solid State Commun. **122**(3–4), 117–120 (2002)
68. A.K.S. Song, R.T. Williams, *Self-Trapped Excitons* (Springer, Berlin, 1996)
69. B.J. Luff, P.D. Townsend, Cathodoluminescence of synthetic quartz. J. Phys. Condens. Matter. **2**, 8089–8097 (1990)
70. F. Messina, L. Vaccaro, M. Cannas, Generation and excitation of point defects in silica by synchrotron radiation above the absorption edge. Phys. Rev. B Conden. Matter Materi. Phys. **81**, 035212 (2010)
71. S. Ismail-Beigi, S.G. Louie, Self-Trapped excitons in silicon dioxide: mechanism and properties. Phys. Rev. Lett. **95**, 156401 (2005)
72. R.M. Van Ginhoven, H. Jónsson, L. René Corrales, Characterization of exciton self-trapping in amorphous silica. J. Non-Cryst. Solids **352**(23–25), 2589–2595 (2006)
73. R. Saavedra, M. León, P. Martín, D. Jiménez-Rey, R. Vila, S. Girard, A. Boukenter, Y. Querdane, Raman measurements in silica glasses irradiated with energetic ions, in *AIP Conference Proceedings*, vol. 1624 (2014), pp. 118–124
74. R.A. Youngman, L.W. Hobbs, T.E. Mitchell, Radiation damage in oxides electron irradiation damage in MgO. J. Phys. Colloq. **41**(C6), 227–231 (1980)
75. T. Sonoda, H. Abe, C. Kinoshita, H. Naramoto, Formation and growth process of defect clusters in magnesia under ion irradiation. Nucl. Instrum. Methods Phys. Res. Sect. B Beam Interact. Mater. Atoms **127–128**, 176–180 (1997). Ion Beam Modification of Materials
76. C. Kinoshita, K. Hayashi, S. Kitajima, Kinetics of point defects in electron irradiated MgO. Nucl. Instrum. Methods Phys. Res. Sect. B Beam Interact. Mater. Atoms **1**(2–3), 209–218 (1984)
77. G.W. Groves, A. Kelly, Neutron damage in MgO. Philos. Mag. **8**(93), 1437–1454 (1963)
78. A.I. Van Sambeek. Radiation-enhanced diffusion and defect production during ion irradiation of MgO and Al2O3, Ph.D. thesis, University of Illinois, Urbana-Champaign, USA, 1997
79. A.I. Van Sambeek, R.S. Averback, C.P. Flynn, M.H. Yang, W. Jager, Radiation enhanced diffusion in MgO. J. Appl. Phys. **83**(12), 7576–7584 (1998)
80. I.O. Usov, J.A. Valdez, K.E. Sickafus, Temperature dependence of lattice disorder in Ar-irradiated (1 0 0), (1 1 0) and (1 1 1) MgO single crystals. Nucl. Instrum. Methods Phys. Res. Sect. B Beam Interact. Mater. Atoms **269**(3), 288–291 (2011)
81. S. Moll, Y. Zhang, A. Debelle, L. Thomé, J.P. Crocombette, Z. Zihua, J. Jagielski, W.J. Weber, Damage processes in MgO irradiated with medium-energy heavy ions. Acta Mater. **88**, 314–322 (2015)
82. B.P. Uberuaga, R. Smith, A.R. Cleave, G. Henkelman, R.W. Grimes, A.F. Voter, K.E. Sickafus, Dynamical simulations of radiation damage and defect mobility in MgO. Phys. Rev. B Condens. Matter Mater. Phys. **71**, 104102 (2005)
83. C.A. Gilbert, S.D. Kenny, R. Smith, E. Sanville, Ab initio study of point defects in magnesium oxide. Phys. Rev. B Condens. Matter Mater. Phys. **76**, 184103 (2007)
84. J. Mulroue, D.M. Duffy, An Ab initio study of the effect of charge localization on oxygen defect formation and migration energies in magnesium oxide. Proc. R. Soc. Lond. A Math. Phys. Eng. Sci. **467**, 2054–2065 (2011)

85. C. Yang, K.G. Malmqvist, M. Elfman, P. Kristiansson, J. Pallon, A. Sjöland, Ionolumines-
cence and PIXE study of inorganic materials. Nucl. Instrum. Methods Phys. Res. Sect. B
Beam Interact. Mater. Atoms **130**(1–4), 746–750 (1997). Nuclear Microprobe Technology and
Applications

86. C. Yang, K.G. Malmqvist, J.M. Hanchar, R.J. Utui, M. Elfman, P. Kristiansson, J. Pallon, A.
Sjöland, Ionoluminescence combined with PIXE in the nuclear microprobe for the study of
inorganic materials, in *AIP Conference Proceedings*, vol. 392, issue 1 (1997), pp. 735–738

87. A. van Wijngaarden, D.J. Bradley, N.M.A. Finney, The ionoluminescence of MgO and
Zn_2SiO_4:Mn. Can. J. Phys. **43**(12), 2180–2191 (1965)

88. D. Kadri, A. Mokeddem, S. Hamzaoui, Intrinsic defects in UV-irradiated MgO single crystal
detected by thermoluminescence. J. Appl. Sci. **5**(8), 1345–1349 (2005)

89. D. Kadri, S. Hiadsi, S. Hamzaoui, Extrinsic defects in UV-irradiated MgO single crystal
detected by thermoluminescence. J. Appl. Sci. **7**(6), 810–814 (2007)

90. E. Shablonin, Processes of structural defect creation in pure and doped MgO and NaCl single
crystals under condition of low or super high density of electronic excitations, Ph.D. thesis,
University of Tartu, Estonia, 2013

91. W.A. Sibley, J.L. Kolopus, W.C. Mallard, A study of the effect of deformation on the ESR,
luminescence, and absorption of MgO single crystals. Phys. Status Solidi (b) **31**(1), 223–231
(1969)

92. S. Datta, I. Boswarva, D. Holt, SEM cathodoluminescence studies of heat-treated MgO crystals.
J. Phys. Colloq. **41**(C6), 522–525 (1980)

93. A.I. Popov, L. Shirmane, V. Pankratov, A. Lushchik, A. Kotlov, V.E. Serga, L.D. Kulikova, G.
Chikvaidze, J. Zimmermann, Comparative study of the luminescence properties of macro- and
nanocrystalline MgO using synchrotron radiation. Nucl. Instrum. Methods Phys. Res. Sect. B
Beam Interac. Mater. Atoms **310**, 23–26 (2013)

94. V. Skvortsova, L. Trinkler, Luminescence of impyrity and radiation defects in magnesium
oxide irradiated by fast neutrons. Phys. Procedia **2**(2), 567–570 (2009). The 2008 International
Conference on Luminescence and Optical Spectroscopy of Condensed Matter

95. V. Skvortsova, L. Trinkler, The optical properties of magnesium oxide containing transition
metal ions and defects produced by fast neutron irradiation, in *In Advances in Sensors, Signals
and Materials* (2010)

96. C. Martínez-Boubeta, A. Martínez, S. Hernández, P. Pellegrino, A. Antony, J. Bertomeu, L.l.
Balcells, Z. Konstantinovic, B. Martínez, Blue luminescence at room temperature in defective
MgO films. Solid State Commun. **151**(10), 751–753 (2011)

97. M.O. Henry, J.P. Larkin, G.F. Imbusch, Nature of the broadband luminescence center in
MgO:Cr^{3+}. Phys. Rev. B Condens. Matter Mater. Phys. **13**, 1893–1902 (1976)

98. CSIRO luminescence database, http://www.csiro.au/luminescence/

99. C.M. MacRae, N.C. Wilson, Luminescence database I-minerals and materials. Microsc.
Microanal. **14**, 184–204 (2008)

Part I
Materials and Methods

Chapter 2
Studied Materials: a-SiO$_2$ and MgO

In this thesis two materials have been studied: amorphous silica (a-SiO$_2$) and magnesium oxide (MgO). Although both a-SiO$_2$ and MgO are oxide materials, their structure is very different: a-SiO$_2$ is amorphous and has a pronounced covalent character, while MgO is crystalline and has a very ionic character. This chapter deals with the structure of both materials and with the characteristics of the samples used in this thesis.

2.1 Amorphous Silica (a-SiO$_2$)

Silicon dioxide (SiO$_2$) can be found in nature either in crystalline (quartz) or in amorphous form (fused silica, a-SiO$_2$). Here we will focus on the properties and the main point defects that can be found in fused silica, although some of these characteristics can also be applied to quartz. The particular properties of the silica samples used in this thesis are explained in this section as well.

2.1.1 Structure of Amorphous SiO$_2$

Zachariasen stated in [1] that: "The principal difference between a crystal network and a glass network is the presence of symmetry and periodicity in the former and the absence of periodicity and symmetry in the latter". He considers that a glass is a continuous random network where there are not two structurally equivalent atoms. Warren confirmed this statement interpreting the X-ray diffraction patterns of vitreous silica using Fourier analysis [2]. The Zachariasen–Warren model is summarized by Galeener in [3]. This model establishes that, although silica does not have a long

© Springer Nature Switzerland AG 2018   17
D. Bachiller Perea, *Ion-Irradiation-Induced Damage in Nuclear Materials*,
Springer Theses, https://doi.org/10.1007/978-3-030-00407-1_2

Fig. 2.1 Representation of
two silica tetrahedra and the
angles θ and ϕ between the
bonds

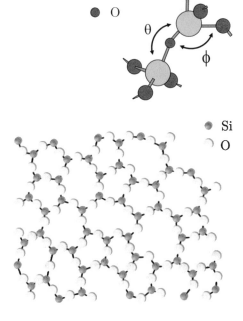

Fig. 2.2 Two-dimensional
representation of a silica
glass network. The rings can
have between 3 and 10
members

range order (neither morphological nor crystalline), it has a short range order with
the following features:

- There are only $Si - O$ bonds. (This happens in both silica and quartz).
- The basic structural units in SiO_2 are tetrahedra (Fig. 2.1) with a Si atom at the
  center and four O atoms, one at each corner of the tetrahedra [3–5]. That means that
  all Si sites are four-fold coordinated and all oxygen sites are two-fold coordinated
  [6]. (This happens in both silica and quartz).
- There is a continuous unimodal distribution of bond lengths peaked at 1.61Å.
  (Only in silica because in crystals this distribution is discrete).
- There is a continuous unimodal distribution of $O - Si - O$ angles (ϕ) peaked at
  109.5°. (Only in silica because in crystals this distribution is discrete).
- There is a continuous unimodal distribution of $Si - O - Si$ angles (θ) peaked at
  144°. (Only in silica because in crystals this distribution is discrete).
- The bond lengths, ϕ and θ are uncorrelated. (Only in silica).
- In silica, atoms form rings between 3 and 10 members (5 and 6-membered rings
  being the most common ones) [7–10], while in quartz only 6 and 8-membered rings
  exist [4]. A two-dimensional scheme of the rings in silica is shown in Fig. 2.2.

The Zachariasen–Warren model also considers that there is a global range order
in silica, because the material is chemically ordered, the structural parameters are
homogeneous statistically, and the macroscopic density predicted by the model is
that of a real silica sample ($\rho = 2.21$ g/cm^3).

The structure of silica can be affected by defects that are classified into two categories: (i) point defects if they only affect the lattice in an isolated site, in the case of a-SiO$_2$ point defects only affect one or two (when an oxygen atom shared by two tetrahedra is involved) silica tetrahedra; (ii) extended defects when their properties are determined by interactions between three or more tetrahedra [11]. Point defects in silica have been studied in this thesis by means of the ion beam induced luminescence technique (Sect. 5.1). Point defects can be produced by irradiation with photons or particles and they affect the optical properties of silica; the different types of points defects in pure silica are reviewed in the next section.

2.1.2 Point Defects in Amorphous SiO₂

The study and understanding of radiation-induced point defects (color centers) in silica is of prime importance since they affect the optical properties of the glass and they are the most important parameter for its applications. For example, point defects affect the optical transparency of silica because they are responsible for several optical absorption bands [9, 12]. This section deals with the different point defects that can be found in pure silica or silica containing OH impurities, because these are the types of silica studied in this thesis. Other defects can appear when the silica glass is doped with other elements such as B, Al, Ge, P or Sn.

Point defects can be paramagnetic or diamagnetic. Paramagnetic defects possess an unpaired electron (dangling bonds) and they can be detected and studied by Electron Paramagnetic Resonance (EPR) spectroscopy (also named Electron Spin Resonance spectroscopy, ESR). In diamagnetic defects there are no unpaired electrons: all the orbitals of the atoms involved in the defect contain two electrons with opposite spins, and these type of defects cannot be detected by EPR (or ESR) spectroscopy.

The paramagnetic defects (Fig. 2.3) that can be produced in pure silica are:

- Self-Trapped Holes (STHs): they result from trapping of holes at neutral oxygen vacancies in the silica network. STHs are only stable at cryogenic temperatures [13] and in low-OH silica (in high-OH silica the STHs are instantly quenched by reaction with hydrogen atoms produced by the radiolysis of the OH groups [14]). There are two types of self-trapped holes [14, 15]:

  - STH$_1$: hole trapped on a single bridging oxygen (see Fig. 2.3a).
  - STH$_2$: hole delocalized over two bridging oxygen atoms of the same SiO$_4$ tetrahedron (see Fig. 2.3b).

- E' centers: there are many different variants of the E' centers (more than 10 different types have been found), but the common feature among all of them is an unpaired electron in a sp^3-like orbital of a 3-fold-coordinated Si atom (a dangling Si bond). They are represented as $\equiv Si\cdot$, where the triple line represents a triple bond and the dot represents an unpaired electron (see Fig. 2.3c).

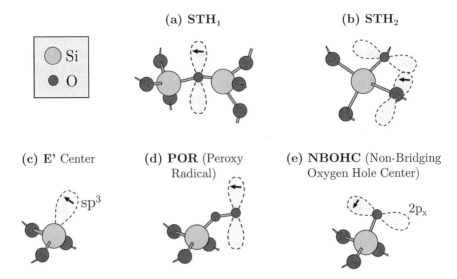

Fig. 2.3 Paramagnetic defects in silica

- Peroxy Radicals (PORs): one oxygen atom in a SiO_4 tetrahedron is bonded to another oxygen atom with an unpaired electron. They are represented as $\equiv Si - O - O\cdot$ (see Fig. 2.3d).
- Non-Bridging Oxygen Hole Centers (NBOHCs): there is an oxygen atom in a silica tetrahedron having one of its three 2p orbitals bonded to a Si atom, two paired electrons in another 2p orbital, and an unpaired electron in the third 2p orbital (a dangling oxygen bond). They are represented as $\equiv Si - O\cdot$ (see Fig. 2.3e).

The diamagnetic defects (Fig. 2.4) that can be found in pure silica or silica containing OH impurities are:

- Peroxy Linkages (POL): between two Si atoms of two tetrahedra there are two oxygen atoms instead of one. This defect is also known as peroxy bridge or interstitial oxygen. They are represented as $\equiv Si - O - O - Si \equiv$ (see Fig. 2.4a).
- Silanol Groups: they are also named hydroxyl groups. They are present in silica containing OH impurities. They are represented as $\equiv Si - O - H$ (see Fig. 2.4b).
- Oxygen Deficient Centers (ODCs): one or two oxygen atoms are missing in a SiO_4 tetrahedron. There are two types of ODCs:
  - ODC(I): electrically neutral relaxed oxygen vacancy. It is represented as $\equiv Si - Si \equiv$ (see Fig. 2.4c).
  - ODC(II): there is a controversy in the literature about the structure of the ODC(II) centers, and two models have been proposed [11, 16]. The most accepted model consists in a divalent Si atom (Si_2^0) [9, 11, 17, 18], also named 2-fold-coordinated silicon. This model is shown in Fig. 2.4d. The second model consists in an electrically neutral unrelaxed oxygen vacancy as shown in Fig. 2.4e [11, 19, 20]. Usually, the notation ODC stands for ODC(II) [11].

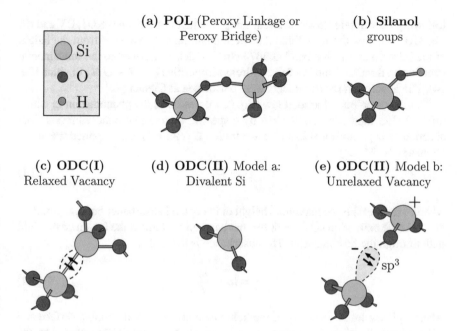

Fig. 2.4 Diamagnetic defects in silica

- Interstitial Molecules: interstitial oxygen molecules (O$_2$, $O - O$) and interstitial ozone molecules (O$_3$, $O - O - O$) can also be present in oxygen-rich silica and in irradiated silica.

The creation of NBOHCs and ODCs (ODC(II)) and their kinetics dependence on the irradiation fluence have been studied in this thesis using the ionoluminescence technique. The behavior under ion irradiation of three silica glasses containing different amounts of OH impurities has been compared.

2.1.3   Calculation of the OH Concentration in Silica from Infrared Spectroscopic Measurements

For the experiments in this thesis, optically polished plates of silica of 6×6 mm^2 area and 1 mm thickness were cut with a diamond disk and cleaned with trichloroethylene and acetone. Three different types of silica have been studied: KU1 from Alkor Technologies (Saint Petersburg, Russia) [21] and two types of silica from Crystran Ltd (Poole, UK) [22], one designed for ultraviolet transmission (UV) and one for infrared transmission (IR). The main difference between the three types of silica is the amount of OH groups that they contain (wet silica having a significant silanol group content and dry silica having a negligible silanol group content). In the following we will refer to the different samples with the nicknames KU1, UV and IR. The OH contents, listed in Table 2.1, were quantitatively assessed from the height of the infrared absorption band at 3673 cm^{-1} which is ascribed to the fundamental stretching vibrational mode of the hydroxyl impurities ($\equiv Si - OH$) in silica [23, 24]. The absorption measurements were performed at Ciemat [25].

The Bouguer-Beer–Lambert law (Eq. (2.1)) describes the phenomenon in which infrared (IR) light can be absorbed by a specific interatomic bond vibration. The absorbance is proportional to the concentration C (mol·l^{-1}) of the bonded species in the material [24, 26].

$$C = \frac{A}{\varepsilon L} \tag{2.1}$$

where A (unitless) is the maximum height of the optical absorbance band, ε (l·mol^{-1}· cm^{-1}) is the extinction coefficient for that band and L (cm) is the length of the light path through the host material. The absorbance is defined as:

$$A = log_{10} \frac{I_0}{I} \tag{2.2}$$

where I_0 is the initial intensity of the light when arriving to the sample surface and I is the intensity of the light after passing through the sample and measured by the spectrophotometer.

In silica glasses there is an absorption band at 3673 cm^{-1} (2.72 μm) in the IR spectrum associated to the hydroxyl impurities present in the material ($\equiv Si - OH$). One can thus calculate the concentration of OH impurities from the height of the peak at 3673 cm^{-1} in the absorbance curves for silica. The OH concentration can be calculated in parts per million (ppm) applying the following conversion to Eq. (2.1) [24]:

$$C_{OH(ppm)} = \frac{A}{\varepsilon L} \cdot \frac{1\,l}{1000\ \text{cm}^3} \cdot \frac{1\ \text{cm}^3}{2.21\ \text{g}} \cdot \frac{17\ \text{g}}{l \cdot \text{mol}_{OH}} \cdot 10^6 \tag{2.3}$$

where we have introduced the value of the density of silica: $\rho = (2.21 \pm 0.01)\,\text{g/cm}^3$.

The value of the extinction coefficient ε (i.e., the molar absorptivity) has been calculated in [24] for silica glass, and it was found to be $(76.4 \pm 2.8)\,l \cdot \text{mol}^{-1} \cdot \text{cm}^{-1}$. The light path through the material is the thickness of the sample: $L = (0.100 \pm 0.005)$ cm.

To determine their OH content of the three types of silica used in this thesis, we have measured their optical absorption in the infrared range. In Fig. 2.5 the measured absorption bands at 3673 cm^{-1} for the three types of samples are shown. Determining the value of the height of the peak (A), the OH contents in the samples have been calculated using Eq. (2.3).

To calculate the peak height we have subtracted a baseline which is different for each sample, so the real height of the peak (A) will be:

$$A = A' - B \tag{2.4}$$

Fig. 2.5 Absorbance spectra of three types of as-received silica samples studied in this work around the 3673 cm^{-1} band

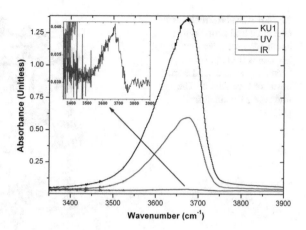

Table 2.1 List of the three types of silica used and their OH content experimentally estimated from their IR absorption spectra

Type of silica	$A \pm \Delta A$ (unitless)	Provided OH content (ppm)	Calculated OH content (ppm)
KU1	1.327±0.056	1000–2000	$(1.34 \pm 0.10) \times 10^3$
UV crystran	0.570±0.023	<1000	(573±42)
IR crystran	0.0132±0.0007	<8	(13±1)

where A' is the value of the maximum of the peak in the original spectra (before subtracting the baseline) and B is the height of the baseline. We obtained the values of A' and B from the spectra in Fig. 2.5. The error in the baseline is 0.001 and the error in A' depends on each spectrum.

We have calculated the overall error of C_{OH} using the following formulae:

$$\varepsilon_A = \varepsilon_{A'} + \varepsilon_B \quad \Rightarrow \quad \Delta A = A \left(\frac{\Delta A'}{A'} + \frac{\Delta B}{B} \right) \tag{2.5}$$

$$\left(\frac{\Delta C_{OH}}{C_{OH}} \right)^2 = \left(\frac{\Delta A}{A} \right)^2 + \left(\frac{\Delta \varepsilon}{\varepsilon} \right)^2 + \left(\frac{\Delta L}{L} \right)^2 + \left(\frac{\Delta \rho}{\rho} \right)^2 \tag{2.6}$$

The results are summarized in Table 2.1 and compared with the OH contents given by the manufacturers. The calculated values are found to be in good agreement with the provided values:

The concentration of metal impurities in our samples is inconsiderable (at the lower ppm level) and they will not affect the luminescence signal [21, 22, 27, 28].

Fig. 2.6 Crystal structure of
MgO: fcc lattice of oxygen
atoms (green circles) with
Mg atoms (red circles)
occupying all the octahedral
holes (or vice versa). The
image was obtained from
[31]

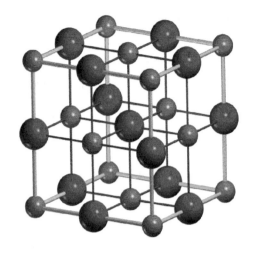

2.2 Magnesium Oxide (MgO)

Magnesium oxide or magnesia (MgO) is a binary, highly ionic compound [29], and
its non-scientific mineral name is periclase. The crystal structure of MgO is a NaCl-
type cubic structure (so-called rock-salt structure), with a face cubic centered (fcc)
lattice of Mg atoms with oxygen atoms occupying all the octahedral holes (or vice
versa), see Fig. 2.6. The Pearson symbol of MgO is $cF8$, which means that the
crystal structure of MgO is cubic, all face centered and contains eight atoms in the
unit cell. MgO belongs to the $Fm\bar{3}m$ space group, it has a bulk lattice parameter of
$a_0 = 1.61\text{Å}$ and a density of $\rho = 3.58 \text{ g/cm}^3$.

The samples used in this study were {100}-oriented MgO single crystals provided by Crystal GmbH, Germany [30]. They were one side epi-polished. The growth
method of the sample was arc melting, the growth direction was [100] and the cleav-
age plane was (100). The size of the samples was $10 \times 10 \text{ mm}^2$ and the thickness was
0.5 mm. The samples were cut into two pieces with a diamond wire and cleaned with
alcohol (ethanol).

2.2.1 Point Defects in MgO

Point defects in crystalline solids as MgO can be intrinsic or extrinsic [32]. Extrinsic
defects are foreign atoms added to the lattice intentionally (solutes) or unintentionally
(impurities), they can occupy lattice sites (substitutional atoms) or interstitial sites.
Intrinsic defects in crystals can be vacancies when an atom is missing in a lattice site,
or interstitials when an atom occupies an interstitial site that would not be occupied
in the perfect crystal. A scheme of vacancies and interstitials is shown in Fig. 2.7.

Intrinsic point defects in crystals can be either Schottky or Frenkel defects [32–
34]. Schottky defects are neutral defects present in ionic and iono-covalent crystals,

Fig. 2.7 Schema of a
vacancy and an interstitial in
a two-dimensional crystal
lattice

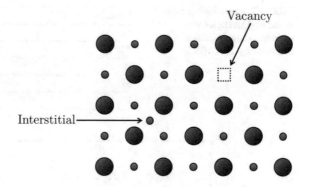

they involve paired vacancies: a cation and an anion are removed from the crystal lattice; in the case of more complicated crystals one vacancy corresponding to each stoichiometric unit is produced. The overall neutral charge of the ionic solid is maintained. On the other hand, Frenkel defects are made up of a paired vacancy and interstitial: an atom of the lattice leaves its place creating a vacancy and stays in a nearby location becoming an interstitial. Both, Schottky and Frenkel defects are represented in Fig. 2.8.

In MgO, oxygen vacancies are known as color centers or F centers (from the word Farbe which means color in German) [34] and Mg vacancies are known as V-centers (cation vacancy) [35]. In MgO also di-interstitials can be present [33].

When MgO is irradiated with electrons, ions, or neutrons at relatively high fluence, clustering of primary defects takes place and extended defects are formed. It is known that $1/2\langle 110\rangle\{100\}$ interstitial dislocation loops are produced in this material under a wide range of irradiation conditions [36–39] and that their growth rate strongly depends on the mass of the ions. The dislocation loop nucleation results from the clustering of point defects.

Fig. 2.8 Schema of a
Schottky defect and a
Frenkel defect in a
two-dimensional crystal
lattice

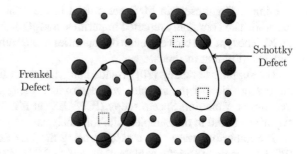

Table 2.2 Impurities present
in our MgO samples and
corresponding quantities
measured by ICP/AES

Impurity	Provided content (ppm)
Ca	198.25
Al$_2$O$_3$	45.32
Fe	47.72
Si	7.11
B	6.89

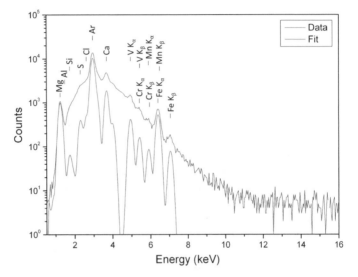

Fig. 2.9 PIXE spectrum of a MgO sample. The blue line corresponds to the experimental spectrum and the red line is the GUPIX simulation

2.2.2   Impurities in the MgO Samples

The impurities present in MgO are crucial in the luminescence emissions of this material. The commonly reported impurities in MgO are: Fe, Cr, Mn, Si, Ti, Al, Ca, Zr, Ni. The type and concentration of impurities in our samples have been determined with two different methods.

The supplier company [30] provided an analysis of the impurities, in this case, the composition of the samples was determined using Inductively Coupled Plasma Atomic Emission Spectroscopy (ICP/AES) at RT. Table 2.2 shows the list of impurities found in our samples with this analysis.

The impurities were also measured using the Particle-Induced X-ray Emission (PIXE) technique (briefly explained in Sect. 3.2.2, p. 43). The measurements were carried out at the Centro Nacional de Aceleradores (CNA) in Seville, Spain [40]. The spectra were taken using an external ion beam in air and two X-ray detectors: Si(Li) and LEGe. The energy of the proton beam used for the experiment was 2.95 MeV when arriving to the surface of the sample (after a 100 nm Si$_3$N$_4$ exit window and 4 mm of air). Figure 2.9 shows one of the spectra recorded with the Si(Li) detector

Table 2.3 Impurities present in our MgO samples and corresponding quantities measured by PIXE with two different detectors. Some of the elements are present in such low concentrations that they are under the limit of detection (<LOD) of the experimental setup

Impurity	PIXE content LEGe (ppm)	PIXE content Si(Li) (ppm)
S	607	422
Ca	413	412
Fe	124	159
Cl	129	<LOD
V	74	91
Mn	<LOD	14
Cr	<LOD	13

(in blue), the simulation performed with the GUPIX program [41] (in red), and some of the elements identified.

The elements identified and their concentration calculated with both detectors are collected in Table 2.3. The concentrations are slightly different than those estimated by the company. One of the main differences observed is the presence of S, Cl, V, and a small quantity of Cr and Mn which is practically under the limit of detection.

References

1. W.H. Zachariasen, The atomic arrangement in glass. J. Am. Chem. Soc. **54**, 3841–3851 (1932)
2. B.E. Warren, H. Krutter, O. Morningstar, Fourier analysis of X-ray patterns of vitreous SiO_2 and B_2O_2. J. Am. Ceram. Soc. **19**(1–12), 202–206 (1936)
3. F.L. Galeener. The physics and technology of amorphous SiO_2, in *Chapter Current Models for Amorphous SiO_2* (Plenum Press, New York, 1987), pp. 1–13. Proceedings of a symposium on the Physics and Technology of Amorphous SiO_2
4. F. Agulló-López, A. Climent-Font, Á. Muñoz-Martín, J. Olivares, A. Zucchiatti, Ion beam modification of dielectric materials in the electronic excitation regime: cumulative and exciton models. Prog. Mater. Sci. **76**, 1–58 (2016)
5. J. Manzano-Santamaría, Daño por excitación electrónica en SiO_2 mediante irradiaciones con iones pesados de alta energía. Ph.d thesis, Universidad Autónoma de Madrid (2013)
6. J. Robertson, The physics and technology of amorphous SiO_2, in *Chapter Electronic Structure of Defects in Amorphous SiO_2* (Plenum Press, New York, 1987), pp. 91–101. Proceedings of a symposium on the Physics and Technology of Amorphous SiO_2
7. F.L. Galeener, R.A. Barrio, E. Martinez, R.J. Elliott, Vibrational decoupling of rings in amorphous solids. Phys. Rev. Lett. **53**, 2429–2432 (1984)
8. A. Pasquarello, R. Car, Identification of Raman defect lines as signatures of ring structures in vitreous silica. Phys. Rev. Lett. **80**, 5145–5147 (1998)
9. L. Skuja, M. Hirano, H. Hosono, K. Kajihara, Defects in oxide glasses. Physica Status Solidi (c) **2**(1), 15–24 (2005)
10. C.E. Jesurum, V. Pulim, L.W. Hobbs, Modeling the cascade amorphization of silicas with local-rules reconstruction. Nucl. Instrum. Methods Phys. Res. Sect. B: Beam Interact. Mater. Atoms **141**(1–4), 25–34 (1998)
11. L. Skuja, Optically active oxygen-deficiency-related centers in amorphous silicon dioxide. J. Non-Cryst. Solids **239**(1–3), 16–48 (1998)
12. L. Skuja, Optical properties of defects in silica, in *Defects in SiO_2 and Related Dielectrics: Science and Technology*. NATO Science Series, vol. 2, ed. by G. Pacchioni, L. Skuja, D.L. Griscom (Springer, Netherlands, 2000), pp. 73–116

13. D.L. Griscom. Self-trapped holes in pure-silica glass: A history of their discovery and characterization and an example of their critical significance to industry. J. Non-Crystall. Solids **352**(23–25), 2601–2617 (2006). Advances in Optical MaterialsAdvances in Optical Materials
14. D.L. Griscom, A minireview of the natures of radiation-induced point defects in pure and doped silica glasses and their visible/near-IR absorption bands, with emphasis on self-trapped holes and how they can be controlled. Phys. Res. Int. **2013**, 379041 (2013)
15. A.V. Kimmel, P.V. Sushko, A.L. Shluger, Structure and spectroscopic properties of trapped holes in silica. J. Non-Crystall. Solids **353**(5–7), 599–604 (2007). Proceedings of the 6th Franco-Italian Symposium on SiO$_2$, Advanced Dielectrics and Related Devices
16. D.L. Griscom, Trapped-electron centers in pure and doped glassy silica: a review and synthesis. J. Non-Crystall. Solids **357**(8–9), 1945–1962 (2011). SiO$_2$, Advanced Dielectrics and Related Devices
17. L.N. Skuja, A.N. Streletsky, A.B. Pakovich, A new intrinsic defect in amorphous SiO$_2$: twofold coordinated silicon. Solid State Commun. **50**(12), 1069–1072 (1984)
18. A.N. Trukhin, Luminescence of localized states in silicon dioxide glass. A short review. J. Non-Crystall. Solids **357**(8–9), 1931–1940 (2011). SiO2, Advanced Dielectrics and Related Devices
19. H. Imai, K. Arai, H. Imagawa, H. Hosono, Y. Abe, Two types of oxygen-deficient centers in synthetic silica glass. Phys. Rev. B: Condens. Matter Mater. Phys. **38**, 12772–12775 (1988)
20. G. Pacchioni, G. Ierano, Computed optical absorption and photoluminescence spectra of neutral oxygen vacancies in α-quartz. Phys. Rev. Lett. **79**, 753–756 (1997)
21. Alkor Technologies, Russia, http://www.alkor.net/FusedSilica_windows_and_lenses.html
22. Crystran, United Kingdom, http://www.crystran.co.uk/optical-materials/silica-glass-sio2
23. E. Vella, R. Boscaino, Structural disorder and silanol groups content in amorphous $Si\,O_2$. Phys. Rev. B: Condens. Matter Mater. Phys. **79**(8), 085204 (2009)
24. K.M. Davis, A. Agarwal, M. Tomozawa, K. Hirao, Quantitative infrared spectroscopic measurement of hydroxyl concentrations in silica glass. J. Non-Crystall. Solids **203**, 27–36 (1996). Optical and Electrical Propertias of Glasses
25. Ciemat, Centro de Investigaciones Energéticas, Medioambientales y Tecnológicas, Madrid, Spain, http://www.ciemat.es/
26. P.R. Griffiths, J.A. de Haseth, *Fourier Transform Infrared Spectroscopy* (Wiley, New York, 1986)
27. K.Yu. Vukolov, Radiation effects in window materials for ITER diagnostics. Fus. Eng. Design **84**(7–11), 1961–1963 (2009). Proceeding of the 25th Symposium on Fusion Technology (SOFT-25)
28. Crystaltechno, Russia, www.crystaltechno.com/FS_UV_en.htm
29. R.W. Davidge, *Mechanical Behavior of Ceramics*, Cambridge Solid State Science Series (Cambridge University Press, Cambridge, 1979)
30. CRYSTAL GmbH, Berlin, Germany, http://www.crystal-gmbh.com
31. http://www.theochem.unito.it/crystal_tuto/mssc2008_cd/tutorials/geometry/geom_tut.html
32. J.W. Morris, Jr. Mater. Sci. (Class Notes), http://www.mse.berkeley.edu/groups/morris/MSE205/Extras/defects.pdf
33. C.A. Gilbert, S.D. Kenny, R. Smith, E. Sanville, Ab initio study of point defects in magnesium oxide. Phys. Rev. B: Condens. Matter Mater. Phys. **76**, 184103 (2007)
34. G. Pacchioni, Ab initio theory of point defects in oxide materials: structure, properties, chemical reactivity. Solid State Sci. **2**(2), 161–179 (2000)
35. P. Baranek, G. Pinarello, C. Pisani, R. Dovesi, Ab initio study of the cation vacancy at the surface and in bulk MgO. Phys. Chem. Chem. Phys. **2**, 3893–3901 (2000)
36. H. Matzke, Radiation damage in crystalline insulators, oxides and ceramic nuclear fuels. Radiat. Effects **64**(1–4), 3–33 (1982)
37. R.A. Youngman, L.W. Hobbs, T.E. Mitchell, Radiation damage in oxides electron irradiation damage in MgO. J. Phys. Coll. **41**(C6), 227–231 (1980)
38. T. Sonoda, H. Abe, C. Kinoshita, H. Naramoto, Formation and growth process of defect clusters in magnesia under ion irradiation. Nucl. Instrum. Methods Phys. Res. Sect. B: Beam Interact. Mater. Atoms **127–128**, 176–180 (1997). Ion Beam Modification of Materials

39. C. Kinoshita, K. Hayashi, S. Kitajima, Kinetics of point defects in electron irradiated MgO. Nucl. Instrum. Methods Phys. Res. Sect. B: Beam Interact. Mater. Atoms **1**(2–3), 209–218 (1984)
40. CNA, Centro Nacional de Aceleradores, Seville, Spain, http://acdc.sav.us.es/cna/
41. GUPIX, http://pixe.physics.uoguelph.ca/gupix/main/

Chapter 3
Ion-Solid Interactions and Ion Beam Modification of Materials

Since this thesis deals with the effects of ion irradiation on materials, it is necessary to understand how ions interact with matter, i.e., how the energy of the projectile ions is transferred to the materials, and what are the main effects of the ion impacts on materials. Also, a brief, general view of the different ion beam modifications of materials that can take place depending on the beam energy is given in this section.

3.1 Ion-Solid Interactions

Two main processes can occur when charged particles pass through matter [1]:

(a) Elastic scattering produced by the nuclei of the target material.
(b) Inelastic collisions with the atomic electrons that produce electronic excitations of the atoms (soft collisions) or ionization of the atoms (hard collisions).

Other possible processes than can occur are: nuclear reactions (inelastic interactions with the nuclei), emission of Cherenkov radiation, and bremsstrahlung. But since these processes are extremely rare in comparison to (a) and (b), they are usually ignored when talking about the interactions between ions and matter.

The predominance of elastic nuclear collisions (nuclear regime) or electronic excitations (electronic regime) depends on the mass and energy of the incident ions (and thus, on their velocity) [2]. The two principal consequences of these two interactions are: the incident particles lose their energy or part of it (their energy is transferred to the material), and the incident particles are deflected from their original direction.

Inelastic (electronic) collisions are the main responsible for the energy loss of heavy particles, the quantity of energy lost in each collision is very small and incident ions suffer small angular deflections which are negligible. However, many electronic collisions are produced and the incident ions lose all their energy in short distances.

© Springer Nature Switzerland AG 2018
D. Bachiller Perea, *Ion-Irradiation-Induced Damage in Nuclear Materials*,
Springer Theses, https://doi.org/10.1007/978-3-030-00407-1_3

In this processes, when ionization is produced, very energetic electrons (named δ-rays, delta electrons or knock-on electrons) are emitted and they can cause secondary ionization.

In the case of elastic (nuclear) collisions, very little energy is transferred in comparison to electronic collisions. However, in the nuclear regime, point defects and defect cascades can be produced generating structural modifications such as topological or chemical disorder, swelling, phase transformations or amorphization [3]. When an amount of kinetic energy higher than a minimum value E_d (threshold displacement energy) is transferred in a collision with a target atom, the knocked atom is ejected from its lattice position generating a Frenkel pair (a vacancy and an interstitial, see Sect. 2.2). If the transferred energy is not very high, the atom will probably return to its site; but if the transferred kinetic energy is high enough, the atom will not recombine with the vacancy, and moreover, the ejected atom will be able to produce secondary displacements and additional vacancies (displacement cascades).

The number of displacements per atom (*dpa*) can be used to measure the quantity of radiation damage produced in irradiated materials [4]. The value of *dpa* is the average number of times that each atom in the material has been displaced from its original site in the lattice. The number of displacements per atom is related to the number of Frenkel pairs produced in the material:

$$dpa = \frac{1}{N} \sum_i n^i n_F^i \qquad (3.1)$$

where N is the atomic density of the material ($atoms/cm^3$), n^i is the number of particles per interaction channel i, and n_F^i is the number of Frenkel pairs produced per interaction channel i.

3.1.1 Stopping Power and Ion Range

The quantity of energy lost by electronic or nuclear processes per unit length during the trajectory is given by the so-called *stopping power*:

$$S_{e,n} = -\left(\frac{dE}{dx}\right)_{e,n} \qquad (3.2)$$

where E is the ion energy, x is the penetration of the ion and the subscripts e and n refer to electronic and nuclear processes, respectively. The total energy loss per unit length is $S = S_e + S_n$. In fact, S_e and S_n are forces and not powers since they are calculated as energy divided by distance ($1\,N = 1\,J/m$), but they are always called *stopping powers* instead of *stopping forces*, and this is the nomenclature that will be used in this thesis to be in accordance with the literature; more information about the origin of this nomenclature can be found in [5]. The minus sign has been introduced

Fig. 3.1 Electronic (S_e), nuclear (S_n) and total ($S_T = S_e + S_n$) stopping powers of bromine ions in silica as a function of the ion velocity. At low velocities (i.e., low energies) the nuclear regime dominates, while at high velocities the electronic collisions are dominant. Two different regions take place in the electronic regime

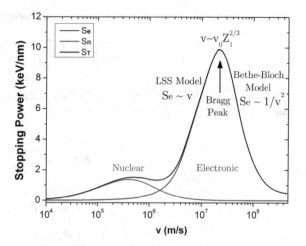

in Eq. (3.2) because dE/dx is a negative value (since it represents the energy loss per unit length) but the S_e and S_n values are always given as positive quantities.

It is worth mentioning that some authors [6, 7] define the stopping power as:

$$S_{e,n} = -\frac{1}{N}\left(\frac{dE}{dx}\right)_{e,n} \tag{3.3}$$

where N is the atomic density of the target. These authors name the quantity $-dE/dx$ the stopping cross-section. However, in this thesis the definition of stopping power that has been used is the one given in Eq. (3.2).

Figure 3.1 shows the two different regimes (nuclear and electronic) that take place as a function of the ion velocity (v, which related to the ion energy by the expression $E = \frac{1}{2}mv^2$).

To define the limits of the two regimes, the velocity of the ions is compared to the Bohr velocity (v_0) which is the velocity of the electrons in the innermost atomic orbitals of hydrogen atoms according to the Bohr model. This value is $v_0 = \frac{c}{137} = 2.19 \cdot 10^6$ m/s (where c is the speed of light) and it corresponds to an energy of $E = 25$ keV/amu [6]. At low energies, when the velocity is lower than the Bohr velocity ($v < v_0$), the nuclear regime is predominant, except in the case of light ions where the electronic stopping power is always higher than the nuclear stopping power. For velocities $v > v_0$ there is an electronic regime in which the energy is mainly lost by inelastic collisions.

Nuclear Stopping Power

There are different models to calculate the nuclear stopping power function depending on the interatomic potential which is considered, the higher level of accuracy is reached by the Ziegler, Biersack and Littmark (ZBL) model [7]. Numerical techniques need to be used to calculate the nuclear stopping powers given by this model.

The ZBL nuclear stopping power for an ion with energy E is:

$$S_n = - \left(\frac{dE}{dx}\right)_n = \frac{\pi a_U^2 \gamma E N}{\varepsilon} \cdot S_n(\varepsilon) \tag{3.4}$$

where a_U is the universal screening length and ε is the reduced energy. The parameters a_U, γ, ε and $S_n(\varepsilon)$ are given by the expressions:

$$a_U = \frac{0.8854 a_0}{Z_1^{0.23} + Z_2^{0.23}} \tag{3.5}$$

$$\gamma = \frac{4 M_1 M_2}{(M_1 + M_2)^2} \tag{3.6}$$

$$\varepsilon = \frac{32.53 M_2 E}{Z_1 Z_2 (M_1 + M_2) \left(Z_1^{0.23} + Z_2^{0.23}\right)} \tag{3.7}$$

where Z_1 and Z_2 are the atomic numbers of the ion and the target atoms, respectively; M_1 and M_2 are the atomic masses of the ion and the target atoms, respectively; a_0 is the Bohr radius ($a_0 = 5.29 \cdot 10^{-11}$ m), and N is the atomic density of the target.

Introducing this expressions in Eq. (3.4) we obtain:

$$S_n = - \left(\frac{dE}{dx}\right)_n = 0.0482 \pi a_0^2 N \frac{Z_1 Z_2 M_1}{(M_1 + M_2) \left(Z_1^{0.23} + Z_2^{0.23}\right)} \cdot S_n(\varepsilon) \tag{3.8}$$

where:

$$S_n(\varepsilon) = \frac{0.5 ln (1 + 1.1383\varepsilon)}{\varepsilon + 0.01321 \cdot \varepsilon^{0.21226} + 0.19593 \cdot \varepsilon^{0.5}} \tag{3.9}$$

Electronic Stopping Power

There are two different regions in the electronic regime (Fig. 3.1) depending again on the ion velocity which determines the effective charge of the ion (i.e., the projectile's state of ionization). Bohr proposed that the effective charge of the ion can be calculated by the formula:

$$\frac{Z_1^*}{Z_1} = \frac{v}{v_0 Z_1^{2/3}} \tag{3.10}$$

where Z_1^* and Z_1 are the effective charge and the atomic number of the ion, respectively. When $v < v_0 Z_1^{2/3}$ the ion is not fully stripped ($Z_1^* < Z_1$), but when $v > v_0 Z_1^{2/3}$ (very high velocities) all the electrons are removed from the ions. The maximum electronic stopping power appears, approximately, at $v \sim v_0 Z_1^{2/3}$ and it is known as the Bragg peak. Two different models describe the electronic stopping power dependence on the ion velocity (or energy) for each region:

(i) When $v < v_0 Z_1^{2/3}$: the Lindhard, Scharff and Shiott model (LSS model) and the Firsov model describe this regime, considering the formation of quasi-molecules. The electronic stopping power in the regime can be calculated by the Lindhard-Scharff formula:

$$S_e = -\left(\frac{dE}{dx}\right)_e = 8\pi e^2 a_0 N \frac{Z_1^{7/6} Z_2}{\left(Z_1^{2/3} + Z_2^{2/3}\right)^{3/2}} \frac{v}{v_0} \tag{3.11}$$

where Z_1 and Z_2 are the atomic numbers of the ion and the target atoms, respectively; a_0 is the Bohr radius ($a_0 = 5.29 \cdot 10^{-11}$ m), and N is the atomic density of the target.

This formula can also be expressed as a function of the ion energy (E) as:

$$S_e = -\left(\frac{dE}{dx}\right)_e = \frac{8\sqrt{2\pi} e^2 a_0 N}{v_0} \frac{Z_1^{7/6} Z_2}{M_1^{1/2} \left(Z_1^{2/3} + Z_2^{2/3}\right)^{3/2}} E^{1/2} \tag{3.12}$$

where M_1 is the mass of the ion.

From Eqs. (3.11) and (3.12) it can be observed that when $v < v_0 Z_1^{2/3}$ the electronic stopping power has a dependence of $S_e \propto v$ or $S_e \propto \sqrt{E}$ on the ion velocity or energy, respectively.

(ii) When $v > v_0 Z_1^{2/3}$: at high energies all the electrons are stripped from the ion, which becomes a bare nucleus, and a pure Coulombic interaction takes place between the ion (without electrons) and the electrons of the target atoms. The Bethe-Bloch model describes this regime and the electronic stopping power is given by the Bethe-Bloch formula:

$$S_e = -\left(\frac{dE}{dx}\right)_e = \frac{4\pi e^4 N}{m_e} \frac{Z_1^2 Z_2}{v^2} \ln\frac{2m_e v^2}{I} \tag{3.13}$$

where I is the mean ionization energy of the target atoms (the average excitation energy of an electron in the target atoms).

This formula is equivalent to:

$$S_e = -\left(\frac{dE}{dx}\right)_e = \frac{4\pi e^4 N}{m_e} \frac{Z_1^2 Z_2}{v^2} \ln\frac{4E}{I} \tag{3.14}$$

When $v > v_0 Z_1^{2/3}$, the dependence of S_e on the ion velocity or energy is: $S_e \propto \dfrac{1}{v^2}$ or $S_e \propto \dfrac{\ln E}{E}$, respectively. In this case, S_e also depends on the atomic number of incident ion as: $S_e \propto Z_1^2$.

Fig. 3.2 Example of an ion traveling a distance R in the target (path in red). The projected range along the incident direction is R_p

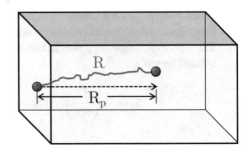

Ion Range

When the ions have lost all their energy, they become implanted in the material. Another important parameter that has to be taken into account when working with ion irradiations is the ion range. The ion range (R) is the total distance traveled by the incident ion in the target material. Since the ion does not travel in a straight path along its incident direction, the projected range (R_p) is defined as the projection of the ion path along its incident direction and corresponds to the penetration of the ion measured from the surface (see Fig. 3.2).

The ion range (R) can be calculated as:

$$R = \int_{E_0}^{0} \frac{1}{dE/dx} dE \tag{3.15}$$

where E_0 is the initial energy of the ion.

Trajectories

The trajectories of the ions in the material depend on their energy range (and thus, on the predominant type of collisions: nuclear or electronic). In nuclear collisions, the ions are highly deflected from their incident trajectory, while in electronic collisions the angular deflections of the ions are negligible. As it is shown for the case of silica in Fig. 3.3a, at low energies (lower than $\sim$100 keV), where the nuclear regime dominates, the ions suffer large random deviations from their initial direction and a large straggling is produced. Figure 3.3b shows the case of high-energy ions (several MeV) in silica; in this case, the electronic regime dominates at the beginning of the trajectories and the ions are not significantly deviated (their trajectories are essentially straight). However, at the end of the path of the ions, when their energy is low and the nuclear collisions dominate, the ions suffer important deviations.

The ions in the electronic regime create amorphous tracks in the target material: cylindrical regions around the ion trajectory where the material is highly defective or amorphous (Fig. 3.4, left). These tracks are produced when the electronic stopping power of the incident ions is higher than a certain threshold value. These value

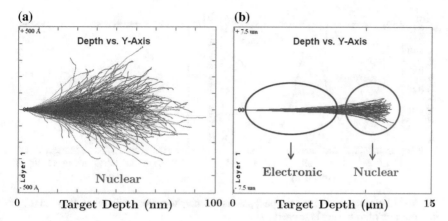

Fig. 3.3 Trajectories of Br ions in silica at **a** 70 keV and **b** 60 MeV. The trajectories were obtained with the program SRIM [8, 9]

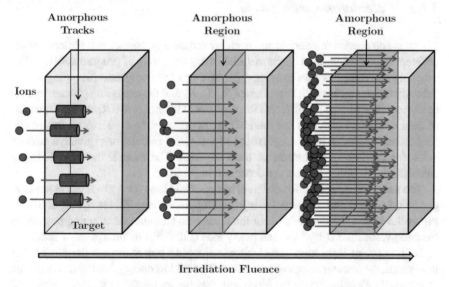

Fig. 3.4 Creation of an amorphous layer by the overlapping of the amorphous tracks when increasing the irradiation fluence

depends on the material, for example, for amorphous silica it has been calculated to be around 2–3 keV/nm and for MgO values between 15–20 keV/nm have been obtained [10]. When the irradiation fluence increases, the amorphous tracks overlap, producing a macroscopic homogeneous defective or amorphous layer in the material [2] (Fig. 3.4, middle). The amorphous layer becomes thicker when increasing the irradiation fluence (Fig. 3.4, right).

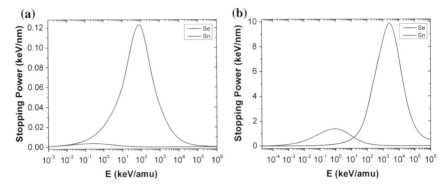

Fig. 3.5 Electronic and nuclear stopping powers in amorphous silica with (**a**) protons and (**b**) Br ions from very low to very high energy

3.1.2 Calculations with SRIM

The stopping powers (electronic and nuclear) and the ion ranges can be calculated for different ions, energies and target materials using the group of programs SRIM (The Stopping and Range of Ions in Matter) [8, 9]. One of the programs (Stopping/Range Tables) provides tables with the values of S_e, S_n, R_p and the straggling (the square root of the variance of R_p) for different ion energies. The program TRIM (the Transport of Ions in Matter) provides the final 3D distribution of the ions and all the kinetic phenomena produced in the target using Monte Carlo calculations (giving a random value for the impact parameter of each particle). The models used by SRIM to calculate the stopping powers are described in [8, 9].

The values of S_e and S_n for different energies and ions for a given target material can be calculated and compared using the Stopping/Range Tables option in SRIM. Figure 3.5 shows the dependence of the electronic and nuclear stopping powers on the energy, for light (Fig. 3.5a) and heavy ions (Fig. 3.5b) in amorphous silica.

In the case of light ions, the electronic stopping power always dominates. For heavy ions, the nuclear stopping power dominates at low energy, having its maximum at around 1 keV/amu, while the electronic regime dominates at high energies. The values of the stopping power are much higher for heavy ions than for light ions (note the different scales in Fig. 3.5a, b). Appendix B shows as an example the input parameters used to obtain the Stopping/Range tables with the SRIM program and the corresponding output text file for gold ions between 1.0 and 1.4 MeV impinging on amorphous silica.

TRIM allows to calculate the stopping powers as a function of the depth and the ion distribution in the target. An example of the input parameters for TRIM is shown in Appendix C for 900 keV Au ions in silica (one of the irradiations used for this thesis). The results of this simulation are shown in Fig. 3.6.

Fig. 3.6 TRIM simulation
for 900 keV Au ions in
silica. a Electronic and
nuclear stopping powers as a
function of the depth. b Ion
distribution as a function of
the depth

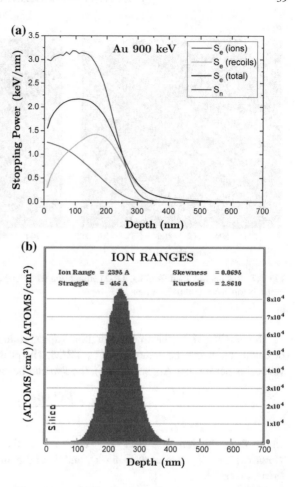

Figure 3.6a represents the electronic and nuclear stopping powers as a function of the depth, for 900 keV Au ions, where the nuclear regime dominates. Most of the ions become implanted right after the maximum energy deposition, being ∼240 nm the mean projected range.

As it can be seen in Fig. 3.6b, the projected range of the ions follows a Gaussian distribution due to the stochastic character of the processes [11]. The Gaussian distribution is centered at the average or mean of the projected range of all the simulated events ($R_p \simeq 240$ nm). The standard deviation of the distribution is the projected range straggling ($\Delta R_p \simeq 46$ nm). The distribution of the ions as a function of the depth (x) and normalized for an implantation fluence of Φ is given by:

$$N(x) = \frac{\Phi}{\Delta R_p \sqrt{2\pi}} \cdot e^{-\frac{1}{2}\left(\frac{x-R_p}{\Delta R_p}\right)^2} \qquad (3.16)$$

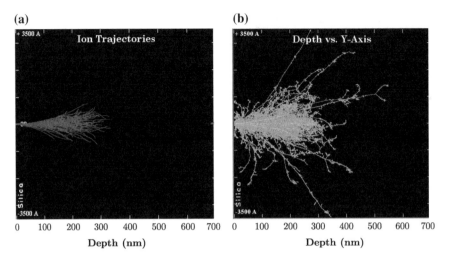

Fig. 3.7 Results obtained from the TRIM simulation with 900 keV Au ions in silica. **a** Ion trajectories. **b** Ion and recoil trajectories

The number of displacements per atom (*dpa*) produced during an irradiation can be calculated using the values given by TRIM. The program provides the number of vacancies per ion per Angstrom (n_V), the *dpa* can be calculated with:

$$dpa = \frac{\phi \cdot t \cdot n_V}{N} \tag{3.17}$$

where ϕ is the ion flux ($ion \cdot s^{-1} \cdot cm^{-2}$), t is the irradiation time (and thus: *Irradiation Fluence* $\equiv \Phi = \phi \cdot t$), and N is the atomic density of the material ($atoms \cdot cm^{-3}$).

TRIM also provides the simulated trajectories of the ions in the material (Fig. 3.7a) and the trajectories of the recoils (Fig. 3.7b). Figure 3.7 corresponds to the simulation used for Fig. 3.6: 900 keV Au ions in silica. It can be seen that the path of almost every impinging ion finishes at around 240 nm.

A new feature of the SRIM 2013 version is the possibility of plotting 3D graphs. Figure 3.8a shows the 3D ion-distribution, where the profile in red corresponds to the ion distribution along the ion beam direction shown in Fig. 3.6b. Figure 3.8b shows the total displacements produced in the silica sample by the 900 keV Au ions.

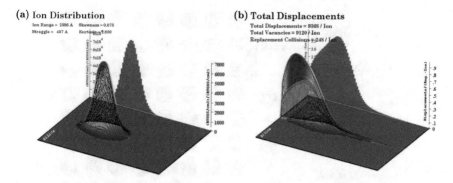

(a) Ion Distribution

Ion Range = 2396 A   Skewness = 0.075
Straggle = 457 A   Kurtosis = 2.850

(b) Total Displacements

Total Displacements = 9368 / Ion
Total Vacancies = 9120 / Ion
Replacement Collisions = 248 / Ion

Fig. 3.8 Results obtained from the TRIM simulation with 900 keV Au ions in silica. **a** 3D ion distribution including a 2D depth plot. **b** 3D and 2D profiles of the total displacements produced by the irradiation

3.2 Ion Beam Modification of Materials and Ion Beam Analysis Techniques

3.2.1 Different Processes of Modification of Materials

Ion irradiation has many applications and different processes occur depending on the energy of the impinging ions. The main processes are summarized in Fig. 3.9.

- Growth of thin films: the technique of growing thin films by bombarding directly with an ion beam on a substrate is known as primary Ion Beam Deposition (IBD) or Low-Energy Ion Bombardment (LEIB). To grow thin films, the energy of the ions must be in the range from a few eV to a few hundred eV; if the beam energy is higher (of the order of a few keV) other processes such as sputtering or implantation can also occur. The ions are produced at the ion source and directed to the substrate by the extractor, which provides the kinetic energy to the ion beam applying an extraction voltage (Fig. 3.10a).
- Sputtering: when the energy of an ion beam impinging on a solid target is of the order of a few keV, the atoms from the surface of the target can be removed and be emitted in every direction. This process is named sputtering and can be used to grow thin films if there is a substrate in front of the solid target (Fig. 3.10b); it can also be utilized to implement etching or analytical techniques. The scheme in Fig. 3.10b shows the configuration when the ion beam is produced by an ion source and an extractor, but the sputtering process can also be produced by particles coming from a plasma, an accelerator or a radioactive material. When this technique is used to fabricate thin films, it is known as secondary ion beam deposition or ion beam sputter deposition [12].
- Ionic implantation: most of the ions become implanted after losing all their energy. This is a very important process used in materials science and materials engineer-

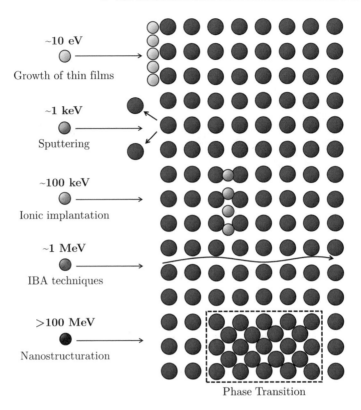

Fig. 3.9 Different processes of modification of materials depending on the beam energy

ing, it is the most used technique to introduce impurities (or dopants) into solids [13, 14]. The main application of ionic implantation is the fabrication of semiconductor devices (doping of the semiconductors). But there are other applications such as changing the physical, chemical or electrical properties of materials, fabricating optical waveguides, etc.

- Ion Beam Analysis (IBA) techniques: when the beam energy is of the order of MeV, these techniques can be used to determine the composition and structure of materials. IBA techniques are described in more detail in the next section, since some of these techniques have been used in these thesis.
- Nanostructuration of materials: relativistic heavy ions can produce phase transitions in materials and can be used to modify the properties of materials at the nanoscale [15, 16]. Two examples of research centers where relativistic ions are produced are the GSI Helmholtz Center in Darmstadt, Germany [17] and the Grand Accélérateur National d'Ions Lourds (GANIL) in Caen, France [18].

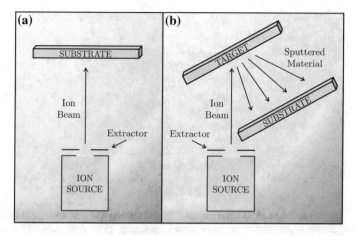

Fig. 3.10 **a** Schema of a primary ion beam deposition setup. **b** Schema of a secondary ion beam deposition (or ion beam sputter deposition) setup

3.2.2 Ion Beam Analysis Techniques

The IBA techniques allow determining the composition and structure of materials. Usually these techniques are non-destructive and can be used in several fields such as archeometry, biomedicine, environment, metallurgy, microelectronics, etc. To use these techniques, the ion beam must have an energy of the order of MeV.

When the ion beam impinges on the target, different processes can occur: the incident ions can be backscattered (if their mass is lower than the mass of the target atoms), the target atoms can be excited by the beam or scattered (if their mass is lower that the mass of the ions), and also nuclear reactions can be produced in which the target nuclei will be converted into different nuclei emitting different particles or photons. There are many different IBA techniques depending on the process studied, the main IBA techniques are briefly described here and summarized in Fig. 3.11.

- RBS (Rutherford Backscattering Spectrometry): the incident ions are elastically scattered by the target nuclei. The backscattered ions are detected with particle detectors. This technique will be explained in more detail in Sect. 5.2. A special configuration of this technique is RBS/C (RBS in channeling mode).
- ERDA (Elastic Recoil Detection Analysis): the target atoms ejected by the impact of the ions from the beam are detected with particle detectors. The mass of the projectile has to be higher than the mass of the target atoms (this technique is usually used to detect light ions). A foil has to be placed in front of the detector to stop the heavy ions from the beam that are scattered by the target.
- PIXE (Particle Induced X-ray Emission): the target atoms are excited or ionized by the projectiles, their subsequent radiative deexcitations produce the emission of X-rays, which are characteristic of each element. The PIXE technique consists of detecting these X-rays.

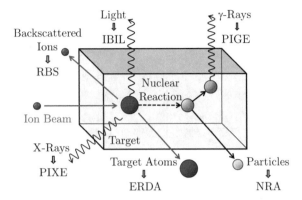

Fig. 3.11 Schema of the main IBA techniques. The red circles represent the ions from the beam, the big blue circles correspond to the target atoms, the orange circle represents a final nucleus in an excited state after a nuclear reaction, this nucleus is represented in green when it has decayed to the ground state, and the small blue circle corresponds to a particle emitted in a nuclear reaction

- PIGE (Particle Induced γ-ray Emission): when a nuclear reaction happens, a target nucleus transforms into a different nucleus in an excited state. This final nucleus falls down to the ground state by emitting γ-rays. The PIGE technique consists of detecting these γ-rays.
- NRA (Nuclear Reaction Analysis): in nuclear reactions different particles can be produced (e.g., protons, α particles,...). NRA is the technique in which these emitted particles are detected.
- IBIL (Ion Beam Induced Luminescence): ionoluminescence is the technique that studies the light emitted by the materials when they are irradiated with ions. This technique will be explained in more detail in Sect. 5.1.

References

1. W.R. Leo, *Techniques for Nuclear and Particle Physics Experiments* (Springer, Berlin, 1987)
2. F. Agulló-López, A. Climent-Font, Á. Muñoz-Martín, J. Olivares, A. Zucchiatti, Ion beam modification of dielectric materials in the electronic excitation regime: cumulative and exciton models. Prog. Mater. Sci. **76**, 1–58 (2016)
3. L. Thomé, A. Debelle, F. Garrido, S. Mylonas, B. Décamps, C. Bachelet, G. Sattonnay, S. Moll, S. Pellegrino, S. Miro, P. Trocellier, Y. Serruys, G. Velisa, C. Grygiel, I. Monnet, M. Toulemonde, P. Simon, J. Jagielski, I. Jozwik-Biala, L. Nowicki, M. Behar, W.J. Weber, Y. Zhang, M. Backman, K. Nordlund, F. Djurabekova. Radiation effects in nuclear materials: role of nuclear and electronic energy losses and their synergy. Nucl. Instrum. Methods Phys. Res. Sect. B: Beam Interact. Mater. Atoms, **307**, 43–48 (2013). The 18th International Conference on Ion Beam Modifications of Materials (IBMM2012)
4. V. Vlachoudis, dpa for FLUKA. http://info-fluka-discussion.web.cern.ch/info-fluka-discussion/lectures/Vlachoudis_DPA_271108.pdf
5. M. Inokuti. http://www.osti.gov/scitech/servlets/purl/71685

6. E. Rauhala, *Instrumental Multi-Element Chemical Analysis, Chapter Scattering Methods* (Springer Science+Business Media B.V, Netherlands, 1998)
7. M. Nastasi, J.W. Mayer, J.K. Hirvonen, *IOn-solid Interactions Fundamentals and Applications*, Cambridge solid state science series (Cambridge University Press, UK, 1996)
8. J.F. Ziegler, J.P. Biersack, U. Littmark, *The Stopping and Range of Ions in Solids* (Pergamon, New York, 1985). http://www.srim.org
9. J.F. Ziegler. Srim, *The stopping and Range of Ions in Matter*. http://www.srim.org/
10. N. Itoh, D.M. Duffy, S. Khakshouri, A.M. Stoneham, Making tracks: electronic excitation roles in forming swift heavy ion tracks. J. Phys. Cond. Matter **21**, 474205 (2009)
11. M. Nastasi, J.W. Mayer, *Ion Implantation and Synthesis of Materials* (Springer, Germany, 2006)
12. J.L. Vossen, W. Kern (eds.), *Thin Film Processes* (Academic Press, USA, 1978)
13. M. Goorsky (ed.), Ion Implantation. InTech (2012). http://www.intechopen.com/books/ion-implantation
14. J.F. Ziegler (ed.), *Ion Implantation Science and Technology*, 2nd edn. (Academic Press, Inc., USA, 1988)
15. M. Lang, F.X. Zhang, R.C. Ewing, J. Lian, C. Trautmann, Z. Wang, Structural modifications of $Gd_2Zr_{2-x}Ti_xO_7$ pyrochlore induced by swift heavy ions: Disordering and amorphization. J. Mater. Res. **24**, 1322–1334 (2009)
16. J. Zhang, R.C. Lang, M. Ewing, R. Devanathan, W.J. Weber, M. Toulemonde, Nanoscale phase transitions under extreme conditions within an ion track. J. Mater. Res. **25**, 1344–1351 (2010)
17. GSI Helmholtzzentrum für Schwerionenforschung GmbH, Darmstadt, Germany. https://www.gsi.de
18. GANIL, Grand Accélérateur National d'Ions Lourds, Caen, France. http://www.ganil-spiral2.eu/

Chapter 4
Experimental Facilities

Most of the experimental techniques used in this thesis are linked to ion accelerator facilities; in particular, electrostatic ion accelerators delivering ions with energies in the range of 500 keV–50 MeV. In this chapter, a general description of such kind of facilities as well as a succinct picture of the three main facilities where this thesis has been carried out is provided. A more detailed description of the used end-stations is given. A description of the ion beam analysis techniques can be found in Chap. 3, Sect. 3.2.2.

The steps in the generation of an ion beam which is suitable for analytical work are [1]:

1. Production of ions from neutral atoms or molecules.
2. Acceleration of these ions by electric fields.
3. Selection of ions with specific mass and energy.
4. Transport of the ions to the specimen to be analyzed.

Keeping this in mind, ion accelerator facilities basically consist on the following elements:

- Accelerator: it is the core of the facility and the rest of the elements are designed depending on the accelerator type and its applications. Electrostatic ion accelerators are based on static electric fields created when an electrical charge is located at an isolated terminal, generating a difference of voltage that can of up to several tens of megavolts between the terminal and the ground. The associated fields are used to accelerate the ions forming the beam.

  Depending on the number of steps to accelerate the ions, electrostatic accelerators can be classified into:

  - **Single ended**: ions are produced inside the accelerator terminal and repelled by the terminal voltage, V_T. The maximum energy that ions with a charge state of $+q$ can reach for a maximum terminal voltage $+V_T$ is:

© Springer Nature Switzerland AG 2018
D. Bachiller Perea, *Ion-Irradiation-Induced Damage in Nuclear Materials*,
Springer Theses, https://doi.org/10.1007/978-3-030-00407-1_4

$$E = E_{acc} + E_{ext} = q \cdot (V_T + V_{ext}) \tag{4.1}$$

where E_{acc} is the energy that the ions gain in the accelerator tank due to the terminal voltage ($E_{acc} = q \cdot V_T$), and E_{ext} is the energy they get due to extraction voltage applied to the source ($E_{ext} = q \cdot V_{ext}$). If the voltages are expressed in Volts, the energy will be obtained in eV.

- **Tandem**: the ions are accelerated in two steps. In the first step the charge state of the ions is -1, being accelerated towards a positive terminal. Once the beam particles arrive at the terminal, they go through an electron stripper element (typically a stripper gas or a stripper carbon foil) which changes their charge state to $+q$. Then, the ions are accelerated in a second step, repelled by the same terminal voltage. In this case the final energy is given by:

$$E = E_{acc} + E_{ext} = E_1 + E_2 + E_{ext} = (q + 1) \cdot V_T + V_{ext} \tag{4.2}$$

where E_1 is the energy in the first step and E_2, the energy in the second step. For the same terminal voltage, the ions reach a higher energy in a tandem accelerator than in a single ended one.

Electrostatic accelerators can also be classified depending on the system used to produce the terminal voltage, being then most important ones:

- **Cockcroft-Walton multiplier**: this system consists of a ladder of capacitors and diodes that converts a low voltage (AC or pulsing DC) into a high DC voltage.
- **Van de Graaff**: this type of accelerators uses a moving belt to accumulate electric charges producing a high DC voltage.

- Ion sources: to produce the ions to be accelerated. Ions can be generated with atoms obtained from a gas source or a solid target, ionized by different methods and extracted after applying an extraction voltage.
- Beam control elements: including beam steerers, ion mass selectors, beam focusing lenses, charge exchange elements, etc. Many of these components are based on the effects of magnetic and electrostatic fields on charged particles (then ions) that form the beam. This elements are used to guide the beams from the ion sources to the experimental stations and to define the size and shape of the beam across the beamlines and at the interaction point with the sample. The most relevant ones installed in the three facilities used for this thesis are:

  - **Magnetic dipoles** consisting on two electromagnets used to apply a magnetic field, perpendicular to the trajectory of the ions and, thus, deviating their trajectory. At the facilities used for this thesis there are two main dipoles: one between the ion sources and the accelerator (low energy magnet), used for selected the ions to be injected into the accelerator, and another one after the accelerator (high energy magnet), used to select the ions with the desired mass and energy to use at the end-station.
  - **Magnetic or electrostatic quadrupoles,** mainly used to focus the ion beam. In particular, for some of the irradiations carried out for this thesis, electro-

static quadrupoles at the exit of the accelerator have been used to overfocus the accelerated beam, allowing for bigger and more homogeneous irradiation areas.
– **Beam monitoring systems**. Located at different points along the path of the ions, they allow the accelerator operator monitoring different properties of the beam. The most relevant ones for this thesis have been Faraday cups (elements to measure the beam current) and Beam Profile Monitors (BPM, they provide the information about the shape and position of the beam).

- Beamlines: vacuum pipes where the ions travel in from the sources to the end-stations. Typical vacuum level varies from 10^{-3} to 10^{-10} mbar, depending on the needs of each experiment.
- End-stations: experimental vacuum chamber at the end of a beamline where the samples are irradiated or analyzed. Experimental chambers are usually equipped with sample holders, vacuum pumps (also present in the beamlines and/or the accelerator), detectors, viewports, etc.

Electrostatic ion accelerators were originally created for nuclear physics, but nowadays, their main applications are analysis of materials (in many different fields, see Sect. 3.2), irradiation of materials to modify their physical properties, and nuclear astrophysics, given the adequate range of energies that they can cover. In this thesis, they have been used for the production and analysis of damage in materials due to ion irradiation.

The experiments of this thesis have been carried out at the Centro de Micro-Análisis de Materiales (CMAM) in Madrid (Spain), at the Centre de Sciences Nucléaires et de Sciences de la Matière (CSNSM) in Orsay (France), and at the Ion Beam Materials Laboratory (IBML) in Knoxville (Tennessee, USA).

4.1 Centro de Micro-Análisis de Materiales (CMAM)

The Center for Micro-Analysis of Materials (CMAM, [2]) is a research facility of Universidad Autónoma de Madrid (UAM) and it is located 15 km at the North of Madrid, Spain. It was officially inaugurated in 2003.

The main facility of the CMAM is a novel electrostatic ion accelerator, built by the Dutch company High Voltage Engineering Europa B.V. (HVEE, [3]). The maximum terminal voltage of the CMAM accelerator is 5 MV and the terminal voltage is generated by a Cockcroft-Walton multiplier. Besides, accelerator at CMAM is an electrostatic tandem accelerator, i.e., the ions are accelerated in two stages, gaining more energy than in a single-ended accelerator (see Eqs. (4.1) and (4.2)).

The main parts and the beamlines of the CMAM accelerator are shown in Fig. 4.1. There are two ion sources at CMAM:

- Sputtering ion source: ions are extracted from a solid target. A Cesium plasma is used to bombard the target pulling up target atoms. These atoms become negative ions when going through the Cs atoms deposited on the target.

Low Energy   Accelerator High Energy   Nuclear   External   Implantation
  Magnet   Tank   Magnet   Beamline   μ-beam line   Beamline

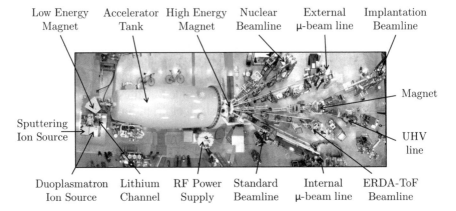

Sputtering
Ion Source

Duoplasmatron   Lithium   RF Power   Standard   Internal   ERDA-ToF
 Ion Source   Channel   Supply   Beamline   μ-beam line   Beamline

Fig. 4.1 Picture of the 5 MV tandem accelerator at the Center for Micro-Analysis of Materials, Madrid. Picture taken by Victor Joco

- Duoplasmatron ion source: ions are extracted from a plasma. This source can be only used for elements in gaseous state. All the ions are extracted with a negative charge state except helium which is extracted as He^+ and then converted to He^- after passing through a lithium channel.

Seven beamlines, with up to 9 end-stations, are available at CMAM, namely: standard beamline, internal μ-beam, ERDA-ToF beamline, surface physics beamline (currently under development), implantation beamline, external μ-beam, PIGE beamline, nuclear physics beamline, and very low temperature irradiation end-station. For this thesis the standard and the implantation beamlines have been used.

The standard beamline is equipped with a multipurpose end-station that can be used for many IBA techniques (IBIL, NRA, ERDA, RBS, RBS/C and PIGE) and for Ion Beam Modification of Materials (IBMM). The chamber is equipped with two particle detectors (one fixed and one movable), a system to put different absorber foils in front of the movable detector, two γ-rays detectors (HPGe and $LaBr_3$), a 4-axis goniometer, a sample holder where several samples can be mounted at the same time, a high sensitivity optical camera, a special viewport for far infrared (thermal) camera, and some extra windows that can be used for other elements such as, for example, optical fibers or lenses (as we did for the ion beam induced luminescence measurements) (Fig. 4.2).

The implantation beamline has recently been developed [4]. This beamline can be used for IBMM, IBA techniques and optical measurements such as ellipsometry, optical absorption, reflectance, etc. The chamber has several viewports which allow to measure different parameters (or to use different techniques) during the irradiations. It allows to perform homogeneous implantations/irradiations over large areas ($10 \times 10\,cm^2$ for 10 MeV protons) by means of a kHz electrostatic scanner. The irradiation chamber is electrically isolated so it can act as a Faraday cup to measure the beam current at any time during the irradiation. The current can also be measured with a system of four independent Faraday cups, which is right before the chamber.

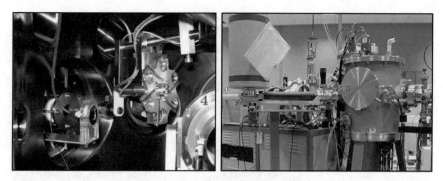

Fig. 4.2 The picture on the left shows the fixed particle detector (1), the movable particle detector (2), the absorber foils (3) and the goniometer (4) in the standard chamber. The second picture shows the standard beamline and a cryostat to refrigerate one of the detectors. Pictures taken by JorgeÁlvarez Echenique

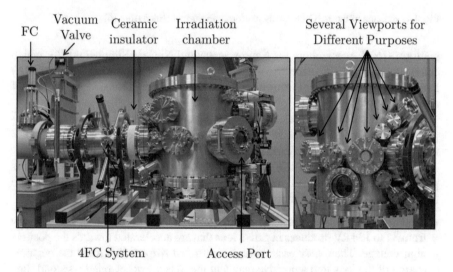

Fig. 4.3 Main elements of the implantation beamline at CMAM

The temperature of the sample can be varied from liquid nitrogen temperature up to 600 °C (Fig. 4.3).

4.2 Centre de Sciences Nucléaires et de Sciences de la Matière (CSNSM)

The Centre de Sciences Nuclèaires et de Sciences de la Matière (CSNSM) is a joint research unit belonging to both CNRS and Universitè Paris-Sud [5]. It is located in Orsay, 30 km at the South-West of Paris, France. It hosts the JANNuS-Orsay

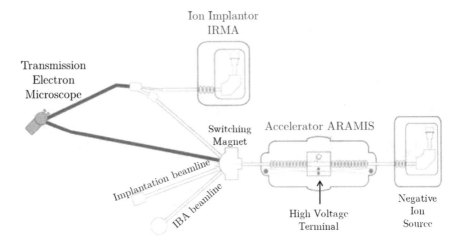

Fig. 4.4 Schema of the JANNuS-Orsay platform. Picture obtained from [6]

platform (Joint Accelerators for Nano-science and Nuclear Simulation) [6], which couples a 2 MV Tandem accelerator (ARAMIS) and a 190 kV ion implanter (IRMA) to a 200 kV Transmission Electron Microscope (TEM FEI Tecnai G^2 20) (Figs. 4.4 and 4.5).

For this thesis the ion accelerator ARAMIS has been used to irradiate MgO samples at elevated temperatures and to study the damage produced in these samples using the RBS/C technique (explained in Sect. 5.2).

ARAMIS is an electrostatic ion accelerator, the high voltage (up to 2 MV) is produced by a Van de Graaff system. The accelerator has two operation modes and two ion sources (one for each mode of operation) [7, 8]:

- Tandem mode: a Cs sputtering ion source with an extraction voltage that can vary from 34 to 134 kV produces negative ions that are accelerated towards the positive high voltage. Then, they pass through a nitrogen stripper that turns the negative charge of the ions into a positive one, and the ions are accelerated a second time as it was explained in Sect. 4.
- Single ended mode: in this mode positive ions are produced by a Penning ion source, placed at the accelerator high voltage terminal. Gaseous elements can be produced with this source. The ions are only accelerated once.

The implantation beamline was used to irradiate and damage the MgO samples with 1.2 MeV Au ions. In this line the beam can be electrostatically scanned over an area up to 12×12 cm^2. The temperature of the samples during the irradiations can be varied from liquid nitrogen temperature up to 900 °C [9]. The sample holder can be tilted 7° to avoid channeling during irradiations. Several samples can be mounted at the same time on the sample holder, a mask is placed in front of the samples to select the sample that has to be irradiated each time. In fact, two diaphragms are used: a large one (L) which is fixed, and a small one (S) that is movable and that is the one used to select the sample. Figure 4.6 shows both diaphragms and the sample.

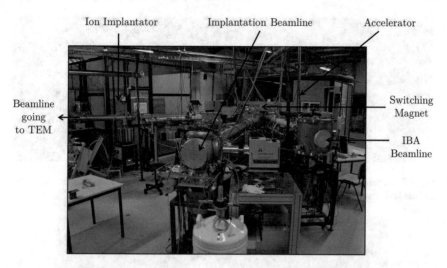

Fig. 4.5 Picture of the CSNSM facilities in Orsay, France. Both implantation and IBA beamlines have been used for this thesis. The third beamline coming from the accelerator (on the left of the picture) is connected to the transmission electron microscope (TEM) where is also connected a beamline coming from the ion implanter

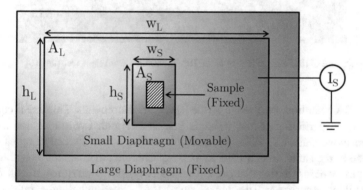

Fig. 4.6 Schema of the sample holder of the implantation beamline at JANNuS-Orsay showing the two diaphragms that are used to delimit the irradiation area and to measure the beam current and thus the irradiation fluence

The area of the big diaphragm is $A_L = w_L \cdot h_L$ and the area of the small diaphragm is $A_s = w_S \cdot h_S$. The beam current is measured at the small diaphragm (I_S), so the beam flux can be calculated as (Fig. 4.6):

$$\phi = \frac{I_S}{q \cdot (A_L - A_S)} \tag{4.3}$$

where q is the charge of the ions in the beam. Once the flux is known, the fluence arriving to the sample can be determined as: $Fluence = \phi \cdot t$, where t is the irradiation time.

Fig. 4.7 Pictures of the irradiation chambers at the CSNSM: the implantation chamber on the left and the IBA chamber on the right

Fig. 4.8 Picture of the ion accelerator at the IBML obtained from their website [10]

The IBA beamline was used to do the RBS/C measurements. This line is equipped with a fixed particle detector, a 4-axis goniometer with two translation and two rotation movements (X, Y, α, β), a sample holder where several samples can be mounted at the same time, a Faraday cup right before the chamber to optimize the beam current before the irradiation, and a chopper system to measure the beam current during all the experiment. The chopper consists of a rotating helix that intercepts the beam reading $1/6$ of the beam current. The current arriving to the sample is thus the value given by the chopper multiplied by five. This system allows to determine the integrated charge (or fluence) during the experiment (Fig. 4.7).

4.3   The Ion Beam Materials Laboratory (IBML)

The Ion Beam Materials Laboratory (IBML, [10]) belongs to the University of Tennessee (UT) and to the Oak Ridge National Laboratory (ORNL). It is located in the UT campus in Knoxville (Tennessee, USA).

The main facility of the IBML is a 3 MV tandem accelerator equipped with two ion sources, three beamlines, and four ion-beam endstations (Fig. 4.8).

Fig. 4.9 Pictures of the sample holders used at room temperature (left) and at low temperature (right)

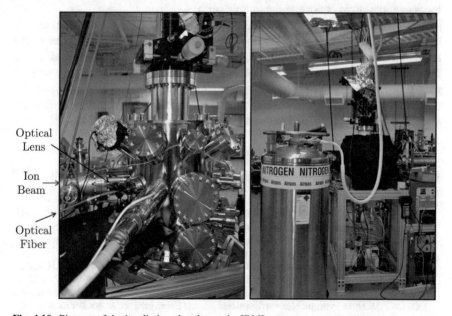

Fig. 4.10 Pictures of the irradiation chamber at the IBML

The experiments of ionoluminescence in silica presented in Sect. 6.3 were carried out at the IBML. All the irradiations were carried out in the same vacuum chamber shown in Figs. 4.9 and 4.10.

Figure 4.9 shows the sample holder for the room temperature (left) and for the low temperature irradiations (right). In both cases the temperature is measured on

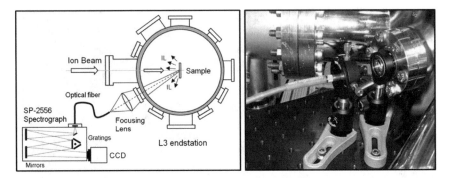

Fig. 4.11 Schema of the IL setup at the IBML (left) and picture of the optical lens and the optical fiber (right). Both figures belong to the IBML. The figure on the left has been obtained from [11, 12]

the surface of the sample holder. Metallic tape and metallic clips are used to avoid the accumulation of charge on the sample surface.

Figure 4.10 shows the ion-beam endstation were the experiments were carried out. During the experiments the chamber is covered with a black fabric to avoid the alteration of the IL spectra by external signals. For the low temperature irradiations, a liquid nitrogen circuit was used to refrigerate. The chamber is equipped with an optical fiber and a spectrometer, an optical lens is used to focus all the light on the optical fiber. The experimental setup used for the ionoluminescence measurements is shown in Fig. 4.11. More details about this setup can be found in [11, 12].

References

1. R.A. Weller, Y.Q. Wang, in *Handbook of Modern Ion Beam Materials Analysis*, 2nd edn. Instrumentation and laboratory practice (Materials Research Society, 2010), pp. 385–424
2. CMAM, Center for micro-analysis of materials, Madrid, Spain, http://www.cmam.uam.es
3. High voltage engineering Europa B.V., http://www.highvolteng.com
4. D. Jiménez-Rey, M. Benedicto, A. Muñoz-Martín, D. Bachiller-Perea, J. Olivares, A. Climent-Font, B. Gómez-Ferrer, A. Rodríguez, J. Narros, A. Maira, J. Álvarez, A. Nakbi, A. Zucchiatti, F. de Aragón, J.M. García, R. Vila, First tests of the ion irradiation and implantation beamline at the CMAM. Nucl. Instrum. Methods Phys. Res. Sect. B Beam Interact. Mater. Atoms **331**, 196–203 (2014). 11th European Conference on Accelerators in Applied Research and Technology
5. CSNSM, Centre de Sciences Nucléaires et de Sciences de la Matière, Orsay, France, http://www.csnsm.in2p3.fr/
6. JANNuS - Joint accelerators for nano-science and nuclear simulation, France, http://jannus.in2p3.fr/spip.php
7. E. Cottereau, J. Camplan, J. Chaumont, R. Meunier, H. Bernas, ARAMIS: an ambidextrous 2 MV accelerator for IBA and MeV implantation. Nucl. Instrum. Methods Phys. Res. Sect. B Beam Interact. Mater. Atoms **45**(1–4), 293–295 (1990)
8. E. Cottereau, J. Camplan, J. Chaumont, R. Meunier, ARAMIS: an accelerator for research on astrophysics, microanalysis and implantation in solids. Mater. Sci. Eng. B **2**(1), 217–221 (1989)

9. H. Bernas, J. Chaumont, E. Cottereau, R. Meunier, A. Traverse, C. Clerc, O. Kaitasov, F. Lalu, D. Le Du, G. Moroy, M. Salomé, Progress report on aramis, the 2 MV tandem at Orsay. Nucl. Instrum. Methods Phys. Res. Sect. B Beam Interact. Mater. Atoms **62**(3), 416–420 (1992)
10. IBML, Ion beam materials laboratory, Knoxville, Tennessee, USA, http://ibml.utk.edu/
11. M.L. Crespillo, J.T. Graham, Y. Zhang, W.J. Weber, In-situ luminescence monitoring of ion-induced damage evolution in SiO_2 and Al_2O_3. J. Luminescence **172**, 208–218 (2016)
12. Y. Zhang, M.L. Crespillo, H. Xue, K. Jin, C.H. Chen, C.L. Fontana, J.T. Graham, W.J. Weber, New ion beam materials laboratory for materials modification and irradiation effects research. Nucl. Instrum. Methods Phys. Res. Sect. B Beam Interact. Mater. Atoms **338**, 19–30 (2014)

Chapter 5
Experimental Characterization Techniques

This chapter describes the three main analysis techniques used in this work: ion beam induced luminescence (ionoluminescence, IL or IBIL), Rutherford Backscattering Spectrometry (RBS), and X-Ray Diffraction (XRD). The three techniques have been used to study the changes and the damage produced in MgO and a-SiO$_2$ by ion irradiation.

RBS/C and XRD are very well-established techniques and detailed explanations about both techniques can be found in the literature [1–10]. However, in order to understand the main principles, the experimental conditions and the nomenclature used in this thesis, a general description of RBS/C and XRD is given here. Both techniques can be only used to characterize crystalline materials, in this thesis RBS/C and XRD were used to characterize the damage produced in MgO by heavy ion irradiation.

Ionoluminescence is less known than RBS/C and XRD, and a large part of this thesis is based on the analysis of the luminescence produced in amorphous silica with different irradiation conditions; therefore, a more detailed description of the technique is offered. Ionoluminescence can be used for both amorphous and crystalline materials, so we have been able to study the ionoluminescence in a-SiO$_2$ and MgO.

5.1 Ion Beam Induced Luminescence (IBIL)

Luminescence is the emission of light by a material not resulting from heat and, particularly, when it is produced by electronic transitions between different energy levels [11]. The three main steps in luminescence are: excitation of the atoms by the absorption of energy, transformation and transfer of the excitation energy, and relaxation to a non-excited state (or less-excited state) producing the emission of light. Luminescence can be intrinsic or extrinsic. In the former case, it is produced

© Springer Nature Switzerland AG 2018
D. Bachiller Perea, *Ion-Irradiation-Induced Damage in Nuclear Materials*,
Springer Theses, https://doi.org/10.1007/978-3-030-00407-1_5

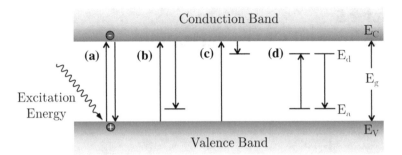

Fig. 5.1 Examples of electronic transitions between the valence and conduction bands. **a** Direct band-to-band transition. **b** Relaxation to an acceptor level (E_a). **c** Relaxation to a donor level (E_d). **d** Excitation and relaxation between an acceptor and a donor level. E_g is the gap energy (around 9 eV in silica), E_C and E_V are the energy levels of the conduction and valence bands, respectively

by direct band-to-band transition (Fig. 5.1a). In the latter case, the recombination takes place via localized states in the forbidden gap (Fig. 5.1b–d). Localized states are discrete energy levels located in the forbidden zone of insulators (between the valence band and the conduction gap) and they are due to the existence of impurities or lattice defects. Extrinsic transitions can be luminescent (a photon is emitted) or non-luminescent (due to the absorption or emission of lattice vibrations). Absorption and emission bands are Gaussian-shaped in terms of the photon energy [12].

Luminescence is a very sensitive technique to identify and investigate optically-active point defects (color centers) in dielectric materials such as SiO_2 [13]. There are several excitation sources to produce luminescence such as electrons (cathodoluminescence, CL), photons (photoluminescence, PL) or X-rays (radioluminescence, RL). The term Ion Beam Induced Luminescence or Ionoluminescence (IBIL or IL) designates the luminescence induced by ion bombardment.

Ionoluminescence offers some advantages in comparison to CL, PL, RL or other alternative methods (electron paramagnetic resonance spectroscopy, optical absorption) to investigate impurities or point defects in materials. The main advantage of IBIL is its intrinsic in situ character which allows for the study of the kinetics of damage (i.e., the dependence of the damage on the irradiation fluence), revealing in real time the structural changes produced in the material by ion irradiation.

Another advantage of IBIL with respect to CL, PL or RL is that tuning ion energy and mass, the depth profile can be modified but also the excitation mechanisms [14]: for relatively slow ions ($E(MeV)/mass(amu) << 1$), the interactions of the ions with the material are mainly elastic (usually called nuclear) collisions; with light ions or with high-energy heavy-mass ions (the so-called swift heavy ions, SHI) deposition of beam energy is primarily due to electronic excitations, i.e., mainly related to the electronic stopping power. Nevertheless, at the end of the ion tracks, there is always a contribution of the nuclear slowing-down; with MeV light ions, this nuclear contribution is very low, but higher nuclear energy deposition rates can be reached using heavy ions at intermediate energies. That means that with the IL technique the rate of energy deposition along the ionization track can be varied, while with CL, PL and RL the stopping power cannot be changed, limiting the possibilities of these techniques. Chapter 7 presents new results on the dependence of ionoluminescence on the stopping power of the impinging ions.

An IL experimental setup requires the following elements:

- A spectrophotometer (or spectrometer) that counts and records the number of photons arriving for each wavelength.
- An optical system (optical fiber, set of lenses,...) to guide the emitted light from the sample to the spectrometer.
- A computer and the software necessary to display and save the spectra registered by the spectrometer.
- A current measurement system: it can be a Faraday cup or a direct current integration system which requires an isolated chamber or sample holder and a secondary-electron suppression voltage.
- A stable and homogeneous beam: the IL signal depends on the beam current, so it is very important to keep the current constant during the measurements.

All the IBIL measurements of this thesis, except for those presented in Sect. 4.3, have been performed at the CMAM [15]. For these experiments two different experimental setups have been used: one for the measurements at room temperature (RT) and the other for measurements at low temperature (80 K).

Samples were irradiated at room temperature (RT) in the standard chamber, at a vacuum of 10^{-6} mbar, connected to the CMAM 5 MV tandem accelerator (see Sect. 4.1). The ion beam was expanded (unfocused) and only the central homogeneous part defined by the slits (4×4 mm^2) was used to irradiate the samples. Fluences (and beam currents) were determined by direct current integration from the target (sample holder) using an electron suppression voltage of +180 V. Since silica and MgO are insulator materials, a graphite tape was used to cover part of the surface of the samples to avoid the electric charge of the surface and the production of electric arcs, which would affect the IBIL spectra and the current measurement (Fig. 5.3, left). This tape did not affect to the size of the irradiation surface, which was delimited by the beam size (always 4×4 mm^2). A sketch and a picture of the experimental setup at RT are shown in Figs. 5.2 and 5.3, respectively.

The irradiations at low temperature were carried out at the implantation beam-line, under a vacuum of 10^{-7} mbar and also connected to the CMAM 5 MV tandem accelerator. In this case, the beam was rastered over an area of 10×10 mm^2, but the irradiation surface was limited and kept constant using a copper mask in front of the samples. The beam current was measured with a Faraday cup before and after the irradiation. A cryostatic system with liquid nitrogen was used to decrease the temperature down to 80 K. A picture of the implantation chamber and of some of the elements of the experimental setup is shown in Fig. 5.4.

In both cases (RT and low temperature) the beam homogeneity, essential for this work, was carefully checked by means of the ionoluminescence induced in a sample of amorphous silica monitored with a 12 bits CCD camera and the size of the beam was determined using a millimeter paper behind this silica sample (Fig. 5.5).

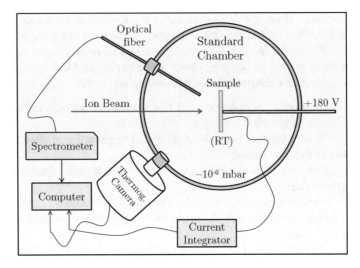

Fig. 5.2 Sketch of the experimental setup used at CMAM to perform the IBIL experiments at room temperature

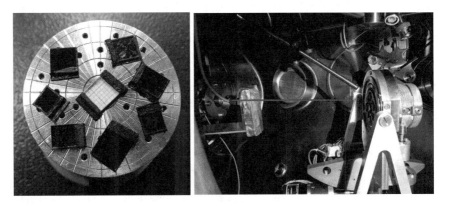

Fig. 5.3 Pictures of the IBIL experimental setup used at CMAM (for the RT measurements). The figure on the left shows the sample holder with the silica samples. The picture on the right shows the sample holder, the optical fiber and the beam direction

Also for both setups, the IL emission was transmitted through a silica optical fiber of 1 mm-diameter. The light was guided to a compact spectrometer QE65000 (Ocean Optics, Inc.) configured with a multichannel array detector for measuring simultaneously the whole spectrum in the range 200–900 nm with a spectral resolution better than 2 nm. The light integration time was 1 s. Absolute values for the IL yield were determined integrating the main peaks of the spectra with the Ocean Optics software: SpectraSuite® [16].

Figure 5.6 shows the working mode of the QE65000 spectrometer [17]. An optical fiber is connected to the SMA (*SubMiniature version A*) connector, guiding the

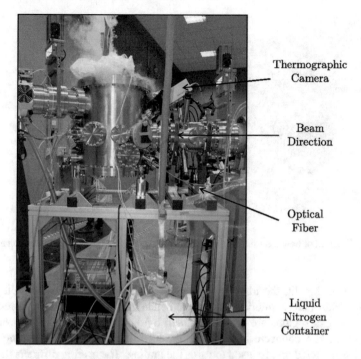

Fig. 5.4 Picture of the implantation chamber at CMAM during the low temperature irradiations

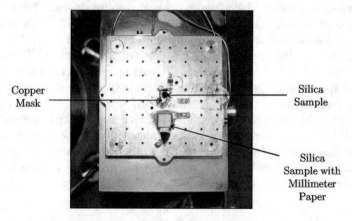

Fig. 5.5 Picture of the sample holder used at the implantation beamline. The upper sample is a silica sample with a copper mask and below there is the sample used to measure the size of the beam and to check its homogeneity

light into the optical bench. Right behind the SMA connector there is a dark piece with a rectangular aperture (slit), the size of the slit determines the amount of light entering into the spectrometer and the spectral resolution. If the spectrometer is

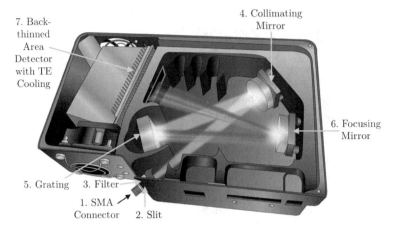

7. Back-
thinned
Area
Detector
with TE
Cooling

4. Collimating
Mirror

6. Focusing
Mirror

5. Grating   3. Filter

1. SMA
Connector   2. Slit

Fig. 5.6 Diagram of how light moves through the optical bench of a QE65000 Spectrometer

used without the slit, the fiber determines the entrance aperture. The light passes through a filter before entering the optical bench, the optical radiation is restricted (the wavelenght regions are pre-determined) by this filter. When the light has past through the SMA connector, the slit and the filter, it goes to the collimating mirror where it is reflected and focused towards the grating. The grating diffracts the light, i.e., it separates the light into different wavelengths. The diffracted light is directed onto the focusing mirror which focuses the light onto the detector. The detector is a Hamamatsu back-thinned (back-illuminated) detector with a 2-D arrangement of pixels (1044 horizontal $\times$ 64 vertical), it can be cooled down to $-15°$ C with the onboard TE-Cooler to reduce dark noise. It provides 90% quantum efficiency.

A common mistake when analyzing IBIL measurements is to fit with Gaussian peaks the spectra as a function of the wavelength [18]. The reason is that the spectrometer provides the spectra as a function of the wavelength, but, as it was mentioned before, absorption and emission bands are Gaussian-shaped in terms of photon energy. This is the reason why a conversion from wavelength to energy (in both axis) has to be done to analyze the IBIL spectra properly. The conversion of the abscissa axis (wavelength) is given by Eq. (5.1) and the conversion of the ordinate axis (yield or number of counts) is given by Eqs. (5.2) and (5.3).

$$E = \frac{hc}{\lambda} \tag{5.1}$$

$$I(\lambda)d\lambda = I(E)dE \tag{5.2}$$

$$I(E) = I(\lambda)\frac{hc}{E^2} = I(\lambda)\frac{\lambda^2}{hc} \tag{5.3}$$

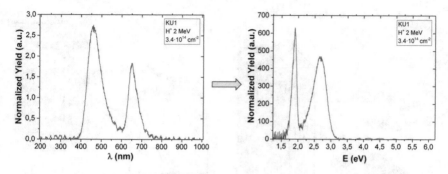

Fig. 5.7 Example of an IBIL spectrum of silica irradiated with 2 MeV protons before and after conversion from wavelength to energy

Figure 5.7 shows an example of an IBIL spectrum of a silica sample during 2 MeV proton irradiation. The original spectrum recorded by the spectrometer is shown on the left of the figure, and the same spectrum after the conversion to energy is on the right. This example shows how important are the changes after conversion, even the position of the maxima can be incorrect if data are treated without this conversion.

5.2 Rutherford Backscattering Spectrometry (RBS)

5.2.1 Description of the RBS Technique

RBS is an ion beam analysis technique that can be used to determine the atomic composition of a sample (elements present in the sample and their atomic concentration, thickness of thin layers, depth profiling, etc.). The depth of the sample that can be analyzed with RBS depends on the mass of the incident ions, their energy and the target composition; for this thesis we have used 1.4 MeV He ions in MgO, allowing to analyze a depth of around 700 nm. Sensitivity, mass resolution and depth resolution (typically 10–30 nm) also depend on the beam selection, the target and the detector; more details can be found in [1, 4]. Although RBS is a very well-known technique and there is a vast amount of information about it in the literature [1–5], a brief description is given here in order to explain and understand some of the results obtained in this thesis.

The principle of RBS is to use an ion beam (in the MeV range) that impinges on the sample that we want to study. Ions penetrate into the target interacting with the atoms and losing a quantity of energy that can be determined knowing the stopping power of the ions. A very small fraction of the ions from the incident beam are backscattered by the atoms in the sample; backscattering occurs after an elastic collision between a projectile and a target nucleus: this collision is described by a Coulomb potential interaction. As the backscattered particle travels back to the

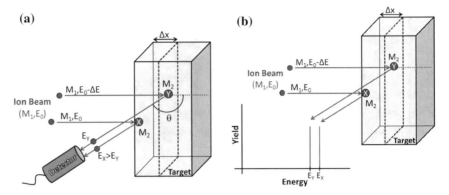

Fig. 5.8 RBS scheme of the backscattering of incident ions on two atoms with the same mass at different depths in the sample. **a** The ions backscattered by the atoms at the surface reach the detector with a higher energy than the ions interacting with atoms located at greater depth. **b** Relative position of the signals in the RBS spectrum corresponding to atoms at different depths

surface, it continues losing energy as occurred when penetrating. A particle detector, located at a certain and well-known angle, is used to detect the backscattered ions; from the processing and analysis of detector signals, a spectrum that represents the number of events as a function of their energy is obtained. Analyzing this spectrum with an appropriate software, we are able to determine the composition of the sample. As it will be explained later, a light-ion beam (H or He) with energy in the MeV range is generally used for RBS analysis, although heavy-ion beams can also be used [1]. A basic scheme of the RBS technique is shown in Fig. 5.8a.

The energy of the ions after the scattering (E') is given by:

$$E' = kE \tag{5.4}$$

where E is the energy of the projectile immediately before the scattering and k is the kinematic factor which depends on three parameters:

$$k = \left[\frac{M_1 cos\theta + \left(M_2^2 - M_1^2 sin^2\theta\right)^{1/2}}{M_1 + M_2} \right]^2 \tag{5.5}$$

M_1 and M_2 are the masses of the incident ions and of the target atoms, respectively, and θ is the scattering angle (Fig. 5.8). The mass of the target atoms (M_2) can be determined since M_1, θ, E and E' are known.

The RBS technique is a particular case of the Backscattering Spectrometry (BS) where the interactions between the incident ions and the target atoms are Coulombic and the cross section of these interactions (the scattering probability) follows the Rutherford formula:

$$\sigma_R\,(E,\theta) = \left[\frac{Z_1 Z_2 e^2}{4E}\right]^2 \frac{4}{sin^4\theta}\frac{\left\{[1-((M_1/M_2)\,sin\theta)^2]^{1/2}+cos\theta\right\}^2}{\left\{1-[(M_1/M_2)\,sin\theta]^2\right\}^{1/2}} \quad (5.6)$$

where Z_1 and Z_2 are the atomic number of the incident ions and of the target atoms, respectively, and E is the energy of the projectile immediately before the scattering. The cross section is essential for the determination of the concentration of the elements in a sample because the backscattering yield is proportional to the scattering probability (Eq. (5.9)). The cross section is higher for lower energies. Although the probability of the interaction is higher for heavier ions (higher Z_1), light ions are usually used for RBS because a higher variety of elements in the samples (light elements) can be detected than using heavy ions, the energy resolution of the detectors is better for light ions and the samples are less damaged. He ions are usually used rather than protons for many reasons: the cross section is higher, interactions with helium have generally Rutherford cross sections at these energies which is not the case for protons, energy loss data are well known for He, and the detector resolution for He is good, about 15 keV or even less in some cases [1].

Figure 5.8 shows what happens with atoms having the same mass but located at different depths in the sample. The incident ions have a mass M_1 which is usually smaller than the mass of the atoms in the sample (M_2). E_0 is the initial energy of the beam. The ions interacting with the nuclei at the sample surface (X) have an energy E_0, while the ions interacting with nuclei located deeper in the sample (Y) have an energy $E_0-\Delta E_{bef}$, where ΔE_{bef} is the energy loss along the distance Δx. The detector only records the ions scattered at an angle θ, and the ions interacting with atoms Y arrive to the detector with less energy (E_Y) than the ions interacting with the atoms at the target surface (E_X) due to the energy loss of the ions passing through the sample before (ΔE_{bef}) and after (ΔE_{af}) being scattered (see Eqs. 5.7 and 5.8). The kinematic factor k is the same in both cases since M_1, M_2 and θ are the same for X and Y. In the RBS spectrum, the count detected after a backscattering event with an atom at the surface will be situated on the right side (higher energy) of the count corresponding to an atom with the same mass but located at a deeper position in the sample (Fig. 5.8b) since $E_X > E_Y$.

$$E_X = kE_0 \quad (5.7)$$

$$E_Y = k\left(E_0 - \Delta E_{bef}\right) - \Delta E_{af} \quad (5.8)$$

One can calculate the depth of the elements in the target (and so the thickness of the layers if a sample is made by different layers) using the stopping power of the incident ions in the target and calculating the total energy loss.

Figure 5.9 shows what happens when having atoms with different masses but at the same depth in the sample (in this example at the surface of the sample). The kinematic factor k is higher for heavier target atoms (higher M_2 in Eq. (5.5)), so the energy of the ions backscattered by heavier ions (E_B) will be higher than the energy of the ions interacting with lighter ions (E_A) as it can be deduced from Eq. (5.4). In

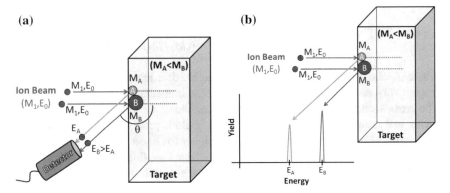

Fig. 5.9 RBS scheme of the backscattering of incident ions on two atoms with different mass at the surface of the sample. **a** The ions backscattered by the heavier atoms reach the detector with a higher energy than the ions interacting with lighter atoms. **b** Relative positions in the RBS spectrum corresponding to elements with a different mass

Fig. 5.10 Schema of the RBS spectrum of a sample having a thin layer composed of two different elements (*A* and *B*, where *B* is heavier than *A*) on the top of a substrate made of a third and lighter element

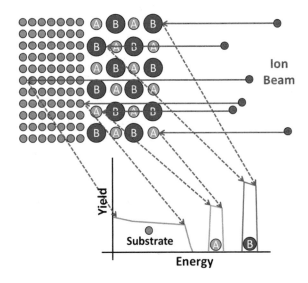

the RBS spectrum, the signal corresponding to heavier ions will appear on the right side of the signal due to light ions (Fig. 5.9b). The cross section will be higher for heavier ions, so the peak corresponding to heavier ions will be higher when having the same concentration of both elements.

A more complicated case with three different elements at different positions in the sample is shown in Fig. 5.10. Thin layers appear as narrow peaks in the spectrum while the substrate of the sample produces a wide signal (because of its thickness).

When the impinging beam is perpendicular to the sample surface, the total number of counts in the peak of the spectrum corresponding to the ith element (the integral of the peak, A_i) is given by:

$$A_i = N_i Q \sigma_i \Omega \tag{5.9}$$

where N_i is the area atomic density (atoms per unit area) of the ith element in the target, Q is the total number of ions arriving to the sample, σ_i is the cross section for the ith element and Ω is the solid angle subtended by the detector.

There are simple cases where some parameters can be easily calculated (i.e., without needing to use a specific software). As an example, if we have a thin film composed of two elements (A and B), the stoichiometry of this layer ($A_a B_b$) can be calculated very easily using Eq. (5.9), even without knowing Q and Ω:

$$\frac{A_A}{A_B} = \frac{N_A Q \sigma_A \Omega}{N_B Q \sigma_B \Omega} \tag{5.10}$$

$$\frac{a}{b} = \frac{A_A \sigma_B}{A_B \sigma_A} \tag{5.11}$$

However, usually the samples are more complex than this example of a thin film composed of two elements, and the RBS analysis requires the use of a specific software. Some examples of programs to analyze RBS spectra (and probably the most known and used by the IBA community) are RBX [19], SIMNRA [20], RUMP [21] and NDF [22–25].

5.2.2 RBS in Channeling Configuration (RBS/C)

A particular configuration of the RBS technique that can be used to analyze crystalline samples consists of aligning the ion beam with one of the major crystal directions of the sample; this configuration is known as channeling, and the RBS technique in channeling mode is usually denoted as RBS/C [1, 4, 5, 26]. This special geometry allows for measuring properties of crystals that cannot be determined with conventional RBS: crystal quality, strain state, dislocations, defects, etc.

In channeling configuration, the ion beam is steered in the channels formed by the rows or planes of atoms in the crystal and the ions penetrate deeper in the material than they would do in the random configuration (Fig. 5.11, left). The channeling mode can also be applied for other techniques such as PIGE/C, PIXE/C or NRA/C. The yield of the RBS/C spectrum will be much lower than the random spectrum (Fig. 5.11, right). The channeling effect is illustrated in Fig. 5.11, where pristine-MgO spectra in random and channeling geometry are shown (red and black symbols, respectively). The two small peaks in the aligned spectrum correspond to the ions backscattered by the atoms at the surface of the sample, where channeling is not yet effective, and they are known as surface peaks. The peak on the right corresponds to the Mg atoms and the peak on the left corresponds to the O atoms (because magnesium is heavier than oxygen).

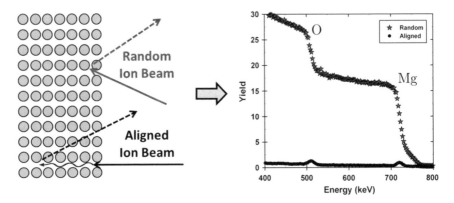

Fig. 5.11 Comparison of a random RBS spectrum and an aligned spectrum of a pristine MgO sample obtained with a 1.4 MeV-He$^+$ beam

Fig. 5.12 Example of a random RBS spectrum and two RBS/C spectra corresponding to irradiated and pristine MgO samples obtained with a 1.4 MeV-He$^+$ beam

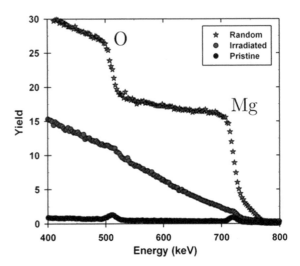

In the case of a damaged sample or if defects are present in the sample, an intermediate spectrum between the random and the aligned spectra in Fig. 5.11 is obtained (Fig. 5.12). The ratio between the yield of an aligned spectrum and the maximum yield (random configuration) provides information about the disorder present in the material.

We used this technique at the CSNSM [27] to characterize the damage produced in MgO crystals irradiated with heavy ions (in particular, 1.2 MeV Au ions). We estimated the damage fraction (the randomly displaced atoms) in MgO for different irradiation fluences performing RBS/C with 1.4 MeV He$^+$ ions and analyzing the spectra. The position of the detector (the scattering angle) for the RBS/C measurements was 165°. We used the Rump program [21] to normalize all the spectra to the

Fig. 5.13 Picture of the
sample holder used for the
RBS/C measurements at the
CSNSM

cumulated charge and solid angle using a rotative random spectrum as a reference
(Fig. 5.13).

The programs mentioned in Sect. 5.2 are suitable for classical (random configuration) RBS analysis but not to analyze RBS/C spectra. However, there are other
programs that allow to do this type of analysis. In the framework of this thesis the
program used for the analysis of the RBS/C spectra is the McChasy code (Monte
Carlo Channeling Simulations) [28, 29]. This program uses Monte Carlo simulations
to reproduce the experimental spectra, which allows determining the defect profile
of the sample (i.e., the damage fraction and dislocations in crystals as a function
of the depth). It is based on the method proposed by J. H. Barret [30] which uses
Thomas–Fermi interactions between ions and lattice atoms.

Although there is a new version of the McChasy code that allows the simulation
of extended defects (such as dislocations), for this thesis the original version of the
program has been used. With this original version one can calculate the defect depth
distribution (defect profile) in the material when using a H or He beam with energy
between 800 and 3500 keV. McChasy splits the crystal into many cells, and only one
cell is taken into consideration. Each cell is divided into a few ten virtual slices, and
the program simulates each slice starting from the surface of the cell and finishing
with the deepest region. The code takes a random initial momentum (this is when
the MonteCarlo method comes in) and calculates the final momentum at the end of
the cell considering the Coulomb screened potential of all the cell atoms. The final
momentum is considered as the initial momentum for the next cell (all the cells are
identical, McChasy always considers the same cell, only the initial momentum is
different). An example of a McChasy input file that has been used for this thesis is
shown in Appendix D. In the input file we have to set all the experimental parameters
(target, axis, ion beam, detector, etc.) and a defect profile that has to be adjusted in
order to fit the simulation results with the experimental spectra by iteration. The
fitting process has to be done from the surface to the deepest regions of the sample
because the cells in depth will be affected by the history of the ion (i.e., what happened

to the ion in the previous cells). More information about the physics and the input protocol of the McChasy code can be found in the user guide of the program and in the references [28, 29, 31, 32].

5.3 X-Ray Diffraction (XRD)

Wilhelm Röntgen discovered X-rays in 1895. X-rays are electromagnetic waves with a wavelength from 0.01 to 10 nm (or an energy in the range of 125 eV–125 keV). This wavelength is on the order of the interatomic distances, making X-rays appropriate to study the atomic structure of crystals using the diffraction phenomenon. Hence, X-ray diffraction (or X-ray crystallography) is a non-destructive technique that allows studying crystalline samples, strain and disorder of crystals.

In order to understand the XRD measurements and the results obtained in this thesis, a concise description of crystalline structures and of the nomenclature and basis of the XRD technique is given here.

5.3.1 Crystalline Structures

A crystalline structure consists in the addition of two elements (Fig. 5.14) [10]:

- A basis: a group of atoms which is identically repeated all along the crystal (Fig. 5.14a).
- A lattice: a set of mathematical points to which the basis is attached (Fig. 5.14b).

A crystal can be defined in the direct (or real) space or in the reciprocal space. The lattice of the crystal in the real space is defined by three translation vectors: $\vec{a}_1, \vec{a}_2, \vec{a}_3$ (known as primitive translation vectors) and the position of every point in the lattice can be described by:

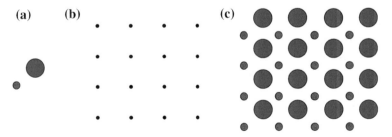

Fig. 5.14 The addition of the basis **a** to the points of the lattice **b** gives rise to a crystalline structure **c**

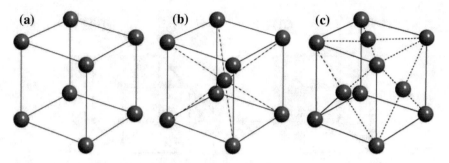

Fig. 5.15 The three types of cubic lattices: **a** sc, **b** bcc, and **c** fcc

$$\vec{r} = \alpha \vec{a}_1 + \beta \vec{a}_2 + \gamma \vec{a}_3 \tag{5.12}$$

where α, β, γ are integers.

The unit cell or primitive cell is the parallelepiped defined by the primitive axis $\vec{a}_1$, $\vec{a}_2$, $\vec{a}_3$ and its volume is given by:

$$V = |\vec{a}_1 \cdot \vec{a}_2 \times \vec{a}_3| \tag{5.13}$$

In three dimensions there are 14 different lattice types (Bravais lattices) classified in seven groups depending on the axis and angles of the cells: triclinic, monoclinic, orthorhombic, tetragonal, cubic, trigonal and hexagonal. There are three cubic lattices: simple cubic (sc), body-centered cubic (bcc), and face-centered cubic (fcc). The crystalline material studied in this thesis (MgO) has a lattice of the fcc type (Fig. 5.15).

Crystalline planes can be defined by the Miller indices, which are a set of numbers that indicates where the plane intercepts the main crystallographic axes of the solid. Miller indices are expressed as (hkl), the values h, k, l corresponding to the inverse of the coordinates where the planes intercept the X, Y and Z axis, respectively. If a Miller index is zero it means that the plane is parallel to the respective axis (the plane intercepts the axis at infinity). A negative value is expressed with a line above the corresponding index.

Some simple examples in a cubic crystal are shown in Fig. 5.16:

The directions and planes of a crystal can be defined by the following notations:

- (hkl): denotes a plane, and h, k, l are the Miller indices of the plane.
- {hkl}: set of all planes equivalent to (hkl) by the symmetry of the lattice.
- [hkl]: denotes a direction vector (in real or direct space).
- $< hkl >$: set of all directions equivalent to [hkl] by symmetry.

A crystal can also be defined in another space which is called the reciprocal space. The reciprocal lattice is defined by the basis $\vec{b}_1$, $\vec{b}_2$, $\vec{b}_3$. The geometric relationship between the direct basis and the reciprocal basis is:

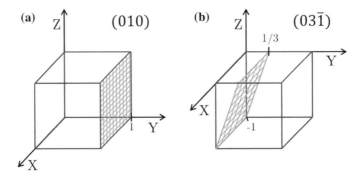

Fig. 5.16 Example of the **a** (010) and **b** (03$\bar{1}$) planes of a cubic crystalline structure

$$\vec{b_1} = \frac{\vec{a_2} \times \vec{a_3}}{V} \quad ; \quad \vec{b_2} = \frac{\vec{a_3} \times \vec{a_1}}{V} \quad ; \quad \vec{b_3} = \frac{\vec{a_1} \times \vec{a_2}}{V} \tag{5.14}$$

which is equivalent to:

$$\vec{a_i} \cdot \vec{b_j} = \delta_{ij} \tag{5.15}$$

where $\delta_{ij} = 1$ if $i = j$ and $\delta_{ij} = 0$ if $i \neq j$. In solid state physics there is a factor 2π in Eqs. (5.14) and (5.15), but the notation here is the one usually used in crystallography.

The reciprocal lattice vector $\vec{\sigma}_{hkl}$ of a family of planes is defined by:

$$\vec{\sigma}_{hkl} = h\vec{b_1} + k\vec{b_2} + l\vec{b_3} \tag{5.16}$$

The Miller indices {hkl} denote the family of planes orthogonal to $\vec{\sigma}_{hkl}$. If the distance between the planes of the family {hkl} is d_{hkl}, the modulus of the reciprocal vector of this family is:

$$|\vec{\sigma}_{hkl}| = \frac{1}{d_{hkl}} \tag{5.17}$$

The use of the reciprocal space is very useful in crystallography and it is particularly useful to understand the X-ray diffraction technique.

5.3.2   Diffraction Phenomenon

In the diffraction phenomenon, X-rays are elastically scattered by the electrons of the atoms in the crystal, producing secondary waves [6, 8]. These waves interfere, but in most directions interferences are destructive. However, in some specific directions the waves may interfere constructively. A scheme of the diffraction phenomenon is

Fig. 5.17 Schema of the X-ray diffraction phenomenon

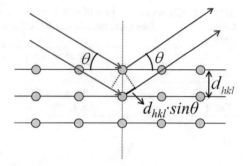

Fig. 5.18 Schema of the X-ray diffraction phenomenon representing the wavevectors and the scattering vector

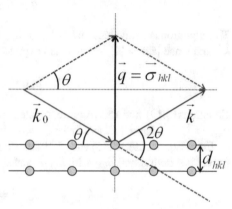

shown in Fig. 5.17. Constructive interferences take place when the Bragg's law is satisfied (Eq. (5.18)).

d_{hkl} is the interplanar distance and θ is the angle between the impinging X-rays and the crystal planes. As it can be deduced from Fig. 5.17 the path difference between two waves coming from two consecutive planes is $2d_{hkl}sin\theta$. Constructive interferences will happen when the secondary waves are in phase and this occurs when the path difference ($2d_{hkl}sin\theta$) is equal to $\lambda, 2\lambda, 3\lambda, ...$ This condition is known as the Bragg's law:

$$2d_{hkl}sin\theta = n\lambda \qquad (5.18)$$

where λ is the X-rays wavelength and n is a natural number. This equation is only satisfied for $\lambda \leq 2d$.

Diffraction can also be understood in terms of the scattering vector ($\vec{q}$) moving in the reciprocal space. If $\vec{k}$ and $\vec{k_0}$ are the scattered and incident wavevectors, respectively, the scattering vector is: $\vec{q} = \vec{k} - \vec{k_0}$. Since the scattering is elastic, the modulus of both $\vec{k}$ and $\vec{k_0}$ is $1/\lambda$ (in Physics a factor 2π is used, but not in crystallography).

The diffraction condition implies that the scattering vector must be equal to the reciprocal lattice vector (Fig. 5.18):

Table 5.1 Conditions on the Miller indices for producing diffraction or not, depending on the crystal lattice. Besides these conditions, the Bragg's rule must also be satisfied for diffraction to occur

Bravais lattice	Possible diffraction	No diffraction
Simple	All	None
Body centered	$h + k + l$ even	$h + k + l$ odd
Face centered	h, k, l unmixed	h, k, l mixed

$$\vec{q} = \vec{\sigma}_{hkl} \tag{5.19}$$

This equation is equivalent to: $\vec{k} - \vec{k}_0 = \vec{\sigma}_{hkl}$. From this we can obtain $(\vec{k}_0 + \vec{\sigma}_{hkl})^2 = \vec{k}^2$, and since $|\vec{k}_0| = |\vec{k}|$, we can also express the diffraction condition as:

$$2\vec{k}_0 \cdot \vec{\sigma}_{hkl} + \sigma_{hkl}^2 = 0 \tag{5.20}$$

Equations (5.19) and (5.20) are equivalent to the Bragg's law (Eq. (5.18)) because $\sigma_{hkl} = 1/d_{hkl}$, $k_0 = 1/\lambda$ and $\vec{k}_0 \cdot \vec{\sigma}_{hkl} = -2k_0\sigma_{hkl}sin\theta$.

From Eqs. (5.15), (5.16) and (5.19) we obtain the Laue equations (doing the scalar product of both $\vec{q}$ and $\vec{\sigma}_{hkl}$ with $\vec{a}_1$, $\vec{a}_2$ and $\vec{a}_3$ in Eq. (5.19)):

$$\vec{a}_1 \cdot \vec{q} = h \quad ; \quad \vec{a}_2 \cdot \vec{q} = k \quad ; \quad \vec{a}_3 \cdot \vec{q} = l \tag{5.21}$$

Another condition must be satisfied to have diffraction: the structure factor (F) must be different from 0. The structure factor is a mathematical expression that describes how a material scatters incident radiation and it is related to each family of planes, taking into account the contribution of every atom in the unit cell. The condition $F \neq 0$ gives rise to a series of rules to easily know for what planes one will be able to observe diffraction depending on the type of crystalline structure of the sample. These rules are summarized in Table 5.1. More details about the scattering theory and the structure factor can be found in [6–10].

Constructive interferences appear in the XRD pattern as a peak (a maximum of intensity) located at the θ value satisfying the Bragg's law for the distance d_{hkl}. When the crystal is perfect, the distance between the planes is always d_{hkl} and there is only one peak in the pattern for each diffraction order. On the contrary, when there are defects in the sample, the crystalline structure can be distorted (Fig. 5.19), which may induce different interplanar distances and thus produce peaks at different values of θ in the pattern (Fig. 5.20). In the case of an inhomogeneously strained crystal the interplanar distances are not constant for all the planes and, in consequence, a broadening of the original peak is produced (Fig. 5.19b). In a homogeneously strained crystal the interplanar distance is constant but different from the pristine sample, thus, a shift of the original peak is produced (Fig. 5.19c).

Figure 5.20 shows two real XRD patterns recorded in the vicinity of the (400) reflection of two {100}-oriented MgO crystals to illustrate the difference between a

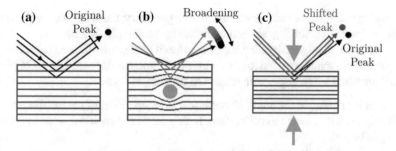

Fig. 5.19 **a** Sketch of the planes of a pristine crystalline sample: the interplanar distance is constant. **b** Inhomogeneously strained crystal: a broadening of the original peak is produced. **c** Homogeneously strained crystal (compressed): a shift of the original peak is produced

Fig. 5.20 Example of the XRD pattern for pristine and irradiated MgO samples

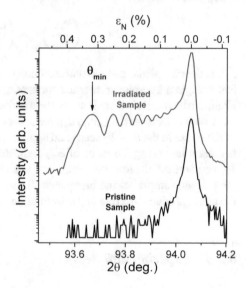

perfect crystal and a distorted one. The black pattern belongs to a pristine MgO single crystal and thus, it shows only one diffraction peak. The red pattern corresponds to a MgO sample irradiated with 1.2 MeV Au ions at 3×10^{13} cm^{-2} at 573 K. The sample has been damaged and the atomic displacements induced changes in the interplanar distance: the pattern contains different peaks that correspond to interferences fringes, this is an evidence of a non-homogeneous strain depth profile [33–36]. The main peak corresponding to d_{hkl} is always present in the patterns because X-rays penetrate deeper in the sample than the region that has been damaged by the Au ions; it corresponds to the intensity diffracted by the unirradiated part of the samples.

There are two frequently used experimental configurations for XRD measurements:

- Symmetric configuration: the incident angle ω is identical to the exit angle θ. Symmetric $\theta - 2\theta$ scans are used to determine the interplanar distance of the planes parallel to the sample surface. In this geometry, the scattering vector $\vec{q}$ and the reciprocal vector $\vec{\sigma}_{hkl}$ are perpendicular to the sample surface.

- Asymmetric configuration: the incident and exit angles are different ($\omega = \theta \pm \alpha$). Asymmetric scans are used to determine the interplanar distance of a set of planes that are tilted by an angle with respect to the sample surface.

The detector measures the angle 2θ because it is independent of the relative position of the planes with respect to the surface. In this thesis symmetric $\theta - 2\theta$ scans have been performed.

Since $\vec{q}$ and $\vec{\sigma}_{hkl}$ are perpendicular to the sample surface in the symmetric configuration, the elastic strain in the direction normal to the surface (ε_N) can be calculated from the $\theta - 2\theta$ scans. The elastic strain of a crystal is the relative variation of the interplanar distance:

$$\varepsilon_N = \left(\frac{\Delta d}{d_{hkl}} \right)_N = \left(\frac{d_{dis} - d_{hkl}}{d_{hkl}} \right)_N \tag{5.22}$$

d_{dis} is the interplanar distance corresponding to a distorted part of the lattice and d_{hkl} is the original interplanar distance corresponding to the pristine sample. The elastic strain can be tensile if there is an increase in the interplanar distance or compressive if the distance decreases. d_{dis} and d_{hkl} (and so ϵ_N) can be calculated using the positions of the peaks in the $\theta - 2\theta$ scans and applying the Bragg's law. The maximum strain is determined using the value of θ_{min} indicated in Fig. 5.20. In this figure it can be seen that the diffraction angle decreases in an irradiated MgO sample with respect to the pristine sample, so the interplanar distance increases (Eq. (5.18)), meaning that the strain produce in MgO under irradiation is tensile (along the surface normal).

5.3.3   Experimental Setup

X-rays can be produced by two different methods [7]:

- Synchrotron-type accelerators: electron are accelerated at very high energy (ultra-relativistic velocities). When these energetic electrons are subjected to a magnetic field producing an acceleration perpendicular to their trajectory, they emit electromagnetic radiation such as X-rays [37–39].
- X-ray tube: a tungsten filament is heated and emits electrons which hit a metallic target where a positive voltage is applied (anode). The atoms of this target are excited by the electrons and then they return to the ground state producing X-rays. The energy of the emitted photons depends on the type of atoms in the target (Cu, Mo, Cr, etc.).

For our measurements we used a X'Pert[3] PRO MRD diffractometer (Fig. 5.21) from PANalytical [40] located at the nanocenter CTU-IEF-Minerve (Universit Paris-Sud, Orsay, France). This diffractometer uses a Cu X-Ray tube. Copper produces X-rays at 1.54 Å (K_α emission), X-rays at 1.39 Å (K_β emission) and bremsstrahlung

Fig. 5.21 Picture of the X'Pert[3] PRO MRD diffractometer from PANalytical

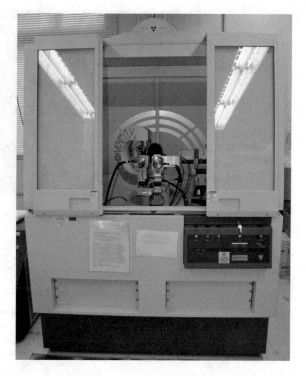

Fig. 5.22 Simplified diagram of the electronic energy levels and transitions of a copper atom (showing only the K lines)

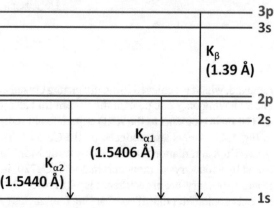

(continuous radiation produced by the deceleration of the electrons when deflected by the atomic nuclei). K_β and bremsstrahlung emissions are removed (usually with a Ni filter or a mirror) and only the K_α emission is used for the diffraction measurements. In fact, there are two contributions for the K_α emission: $K_{\alpha 1}$ at 1.5406 Å and $K_{\alpha 2}$ at 1.5440 Å. The diagram of the Cu lines is shown in Fig. 5.22. As the wavelengths of the $K_{\alpha 1}$ and $K_{\alpha 2}$ emissions are very close, in the case of using High-Resolution X-Ray Diffraction (HRXRD, as in our measurements), the $K_{\alpha 2}$ emission must be

Fig. 5.23 Picture of the X'Pert³ PRO MRD diffractometer from PANalytical showing the main elements

Fig. 5.24 Schema of a
7-axis goniometer

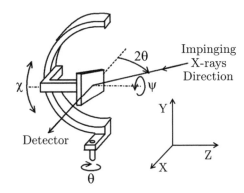

removed, which is achieved using diffracting crystals. A monochromatic X-ray beam allows having only one peak in the pattern for each (hkl) plane.

The main elements of the XRD experimental setup used for this thesis are shown in Fig. 5.23. X-rays are produced by the Cu tube. The monochromatic and parallel beam (Cu $K_{\alpha 1}$ radiation) is obtained by using a multilayer mirror behind the tube followed by a four-crystal monochromator (4xGe220) in asymmetric configuration; the resulting primary-beam divergence is ∼18 arcsec (0.005°). The sample is mounted on a 7-axis goniometer that allows varying four angles (ψ, χ, θ and 2θ) and performing translations in X, Y, Z (Fig. 5.24).

There are two detectors in order to perform measurements with different resolutions. For our measurements a three-bounce crystal analyzer (3xGe220) was placed in front of the detector to improve the angular resolution.

References

1. J.R. Tesmer, M. Nastasi (eds.), *Handbook of Modern Ion Beam Materials Analysis* (Materials Research Society, Pittsburgh, Pennsylvania, USA, 1995)
2. W.K. Chu, J.W. Mayer, M.A. Nicolet, *Backscattering Spectrometry* (Academic, New York, 1978)
3. M. Mayer, in *Rutherford Backscattering Spectrometry (RBS)*. Lectures given at the Workshop on Nuclear Data for Science and Technology: Materials Analysis (2003)
4. A. Redondo-Cubero, Structural and compositional characterization of wide bandgap semiconductor heterostructures by ion beam analysis, Ph.D. thesis, Universidad Autónoma de Madrid, 2010
5. B.N. Dev, Surface and near-surface modification and analysis by mev ions. Curr. Sci. **80**(12), 1550–1559 (2001)
6. A. Debelle, *On the Use of X-ray Diffraction for the Study of Irradiated Materials - Application to Oxide and Carbides* (Université Paris-Sud, Habilitation à Diriger des Recherches, 2013)
7. A. Debelle, *High-Resolution X-ray Diffraction for the Study of Irradiated Single-crystalline Materials* (University of Tennessee, Tennessee, 2015)
8. B.E. Warren, *X-ray Diffraction*, 2nd edn. (Dover Publications, Mineola, 1990)
9. M.A. Krivoglaz, *Theory of X-ray and Thermal-Neutron Scattering by Real Crystals* (Plenum, USA, 1969)
10. C. Kittel, *Introduction to Solid State Physics*, 8th edn. (Wiley, USA, 2005)
11. R. Sahl, in *Crystalline Silicon-Properties and Uses*, Defect related luminescence in silicon dioxide network: a review (InTech, Rijeka, Croatia, 2011), pp. 135–172
12. A.F. Lubchenko, On the shapes of bands of light absorption and emission by impurities. Ukrains'kii Fizichnii Zhurnal **1**(3), 265–280 (1956)
13. R.C. Ropp, *Luminescence and the Solid State* (Elsevier, Amsterdam, 2004)
14. P.D. Townsend, Y. Wang, Defect studies using advances with ion beam excited luminescence. Energy Procedia **41**, 64–79 (2013). International Workshop Energy 2012
15. CMAM, Center for Micro-Analysis of Materials, Madrid, Spain. http://www.cmam.uam.es
16. Ocean Optics, Inc, Dunedin, FL, USA, in *SpectraSuite Spectrometer Operating Software: Installation and Operation Manual*
17. Ocean Optics, Inc, Dunedin, FL, USA, in *QE65000 Scientific-grade Spectrometer: Installation and Operation Manual*
18. Y. Wang, P.D. Townsend, Common mistakes in luminescence analysis. J. Phys. Conf. Ser. **398**(1), 012003 (2012)
19. E. Kótai, *RBX: Simulation of RBS and ERD Spectra* (Research Institute for Particle and Nuclear Physics, Hungary, 1985)
20. M. Mayer, *SIMNRA* (Max-Planck-Institute for Plasma Physics, Germany, 1996)
21. M. Thompson, *RUMP: Rutherford Backscattering Spectroscopy Analysis Package* (Cornell University, USA, 1983). http://www.genplot.com
22. N.P. Barradas, C. Jeynes, R.P. Webb, Simulated annealing analysis of Rutherford backscattering data. Appl. Phys. Lett. **71**, 291–293 (1997)
23. N. Barradas, C. Jeynes, R. Webb, *NDF (Ion Beam Analysis DataFurnace)* (1997). http://www.surrey.ac.uk/ati/ibc/research/ion_beam_analysis/ndf.htm
24. C. Jeynes, N.P. Barradas, P.K. Marriott, G. Boudreault, M. Jenkin, E. Wendler, R.P. Webb, Elemental thin film depth profiles by ion beam analysis using simulated annealing - a new tool. J. Phys. D Appl. Phys. **36**(7), R97–R126 (2003)
25. C. Jeynes, M.J. Bailey, N.J. Bright, M.E. Christopher, G.W. Grime, B.N. Jones, V.V. Palitsin, R.P. Webb, "Total IBA" - where are we? Nucl. Instrum. Methods Phys. Res. Sect. B Beam Interact. Mater. Atoms **271**, 107–118 (2012)
26. T.L. Alford, L.C. Feldman, J.W. Mayer, *Fundamentals of Nanoscale Film Analysis* (Springer, New York, 2007)
27. CSNSM, Centre de Sciences Nucléaires et de Sciences de la Matière, Orsay, France. http://www.csnsm.in2p3.fr/

28. L. Nowicki, A. Turos, R. Ratajczak, A. Stonert, F. Garrido, Modern analysis of ion channeling data by Monte Carlo simulations. Nucl. Instrum. Methods Phys. Res. Sect. B Beam Interact. Mater. Atoms **240**(1–2), 277–282 (2005)
29. L. Nowicki, *McChasy - Monte Carlo Channeling Simulations* (The Andrzej Soltan Insitute for Nuclear Studies, Warsaw, Poland, 2006)
30. J.H. Barrett, Monte carlo channeling calculations. Phys. Rev. B Condens. Matter Mater. Phys. **3**, 1527–1547 (1971)
31. P. Jozwik, Analysis of crystal lattice deformation by ion channeling (2011). http://www.itme. edu.pl/tl_files/Zaklady/Z-2/Seminaria/PJozwik.pdf
32. P. Jozwik, N. Sathish, L. Nowicki, J. Jagielski, A. Turos, L. Kovarik, B. Arey, S. Shutthanandan, W. Jiang, J. Dyczewski, A. Barcz, Analysis of crystal lattice deformation by ion channeling. Acta Phys. Pol. A **123**(5), 828–830 (2013)
33. V.S. Speriosu, Kinematical x-ray diffraction in nonuniform crystalline films: strain and damage distributions in ion-implanted garnets. J. Appl. Phys. **52**, 6094–6103 (1981)
34. S. Rao, B. He, C.R. Houska, X-ray diffraction analysis of concentration and residual stress gradients in nitrogen-implanted niobium and molybdenum. J. Appl. Phys. **69**(12), 8111–8118 (1991)
35. A. Debelle, A. Declémy, XRD investigation of the strain/stress state of ion-irradiated crystals. Nucl. Instrum. Methods Phys. Res. Sect. B Beam Interact. Mater. Atoms **268**(9), 1460–1465 (2010)
36. A. Boulle, A. Debelle, Strain-profile determination in ion-implanted single crystals using generalized simulated annealing. J. Appl. Crystallogr. **43**, 1046–1052 (2010)
37. F.R. Elder, A.M. Gurewitsch, R.V. Langmuir, H.C. Pollock, Radiation from electrons in a synchrotron. Phys. Rev. **71**, 829–830 (1947)
38. D. Iwanenko, I. Pomeranchuk, On the maximal energy attainable in a betatron. Phys. Rev. **65**, 343–343 (1944)
39. V. Veksler, A new method of the acceleration of relativistic particles. Proc. USSR Acad. Sci. **43**, 346 (1944)
40. PANalytical. http://www.panalytical.com/XPert3-MRD-XL.htm

Part II
Ion Beam Induced Luminescence in Amorphous Silica

Chapter 6
General Features of the Ion Beam Induced Luminescence in Amorphous Silica

The typical silica IL spectrum and the origin of the main bands observed on it are described in Sect. 6.1. Two main characteristic bands are always present in the silica IL spectra: a red band peaked at 1.9 eV (650 nm) and a blue band at 2.7 eV (460 nm).

Two important questions when studying IBIL in silica will be also answered in this chapter. The first question is the origin of the blue band, as there is a controversy in the literature when dealing with this problem. Some results of IL in silica at low temperature are reported in Sect. 6.2 to help clarifying this controversy.

The second issue is about the importance of the nuclear stopping power contribution in the IL spectra. Experiments carried out with Au ions in the electronic and in the nuclear regime are presented in Sect. 6.3 to answer this question.

6.1 General Features of the Ionoluminescence Signal in Silica at Room Temperature

Figure 6.1 shows an example of a typical silica IL spectrum obtained with proton irradiation at room temperature. It was measured with the experimental system described in Fig. 5.2 (p. 62). Although, for a better comprehension, some of the spectra shown in this thesis are expressed as a function of the wavelength (λ), the spectrum of Fig. 6.1 has been converted as a function of the energy by using Eq. 5.3 (p. 64) for the Y-axis, as it is explained in [1].

As this figure shows, silica spectra at RT always present two predominant emission bands peaked at 450–460 nm (2.7 eV, blue band) and 650 nm (1.9 eV, red band). Some additional emissions at around 1.8 and 2.3 eV are also observed; they show a smaller intensity and have been usually associated to intrinsic defects, they will not be further considered in this work. The parameters of the main bands, derived from a Gaussian decomposition performed with the program Fityk [2], are provided in Table 6.1.

© Springer Nature Switzerland AG 2018
D. Bachiller Perea, *Ion-Irradiation-Induced Damage in Nuclear Materials*,
Springer Theses, https://doi.org/10.1007/978-3-030-00407-1_6

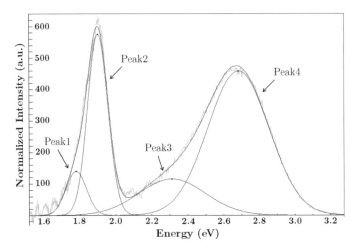

Fig. 6.1 KU1 IL spectrum obtained irradiating with 2 MeV protons at room temperature and at a fluence of 3.4×10^{14} cm^{-2} (green line). The four red curves are the Gaussian functions used for the fitting. The addition of these four peaks results in the final fit represented by the black line

Table 6.1 Parameters of the peaks obtained from the fitting of the curve presented in Fig. 6.1

	Peak 1	Peak 2	Peak 3	Peak 4
Center (eV)	1.78	1.90	2.31	2.68
Center (nm)	697	653	537	463
Height (a.u.)	140	576	115	458
HWHM (eV)	0.067	0.067	0.218	0.201

The two main bands have often been observed under different types of irradiation, although slightly different peak positions and widths are often reported. The red band has been generally associated to Non-Bridging Oxygen Hole (NBOH) centers [3, 4] in base to photoluminescence (PL) and Electronic Paramagnetic Resonance (EPR) experiments. The origin of the blue band is less clear and is, often, a matter of controversy. Section 6.2 is devoted to clarify this controversy and to explain why the origin of the blue band is a luminescence triplet-to-singlet transition of the oxygen-vacancy centers (ODCs). The measured width as well as the kinetic evolution (Chap. 7) obtained in our irradiation experiments at RT, clearly suggest that the blue band, like the red one, should be associated to extrinsic radiation-induced recombination. Specifically, we ascribe it to self-trapped exciton (STE) recombination at irradiation-induced oxygen-deficient centers (ODCII).

In fact, ODCs have two possible transitions [5] that are shown in Fig. 6.2. Using the photoluminescence technique it has been observed [5] a band at 2.7 eV (460 nm) associated to the ODCII triplet-to-singlet transition ($T_1 \rightarrow S_0$) and a band at 4.4 eV (282 nm) associated to the ODCII singlet-to-singlet transition ($S_1 \rightarrow S_0$).

Fig. 6.2 Two possible transitions of the ODCII centers in fused silica

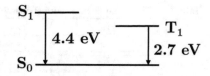

Fig. 6.3 IBIL spectra of the three types of silica showing a small peak at ∼4.4 eV

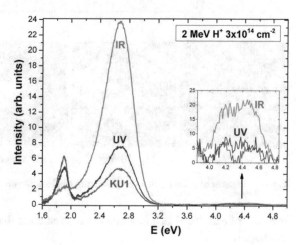

Table 6.2 Position and origin of the three main ionoluminescence emissions of amorphous silica

λ (nm)	E (eV)	Associated defect
∼650 (red)	1.9	NBOHC
∼460 (blue)	2.7	ODCII ($T_1 \rightarrow S_0$)
∼282	4.4	ODCII ($S_1 \rightarrow S_0$)

Indeed, a small peak at ∼4.4 eV can be seen at the high-energy region of the IBIL spectra of silica. Figure 6.3 shows three IBIL spectra obtained with 2 MeV protons, they correspond to the three types of silica studied in this thesis. The existence of this small peak and the results obtained by photoluminescence [5] point out that the blue emission is associated to the ODCII centers.

Table 6.2 summarizes the positions and origins of the peaks shown in Fig. 6.3.

The two main bands of silica (1.9 and 2.7 eV) change during ion irradiation, in some cases this can be observed even with the naked eye. As an example, Fig. 6.4 shows four pictures (corresponding to four different fluences) of a KU1 sample. The pictures were taken during irradiation with 4 MeV He$^+$ ions. At the beginning of the irradiation the red emission is predominant, but when the fluence increases, the blue emission becomes more important. The IR samples show a different behavior: their color is blue during all the irradiation.

The evolution of both bands depends on the type of silica (in particular, on their impurity content) and on the irradiation conditions. It is precisely this evolution (the

Table 6.3 Times and fluences corresponding to the spectra in Fig. 6.6

KU1	(a)	$t = 3$ s	$Fluence = 10^{12}$ cm^{-2}
	(b)	$t = 8$ s	$Fluence = 2.8 \times 10^{12}$ cm^{-2}
IR	(c)	$t = 1$ s	$Fluence = 4 \times 10^{11}$ cm^{-2}
	(d)	$t = 9$ s	$Fluence = 3.3 \times 10^{12}$ cm^{-2}

2.2 · 10^{13} cm^{-2} 2.6 · 10^{14} cm^{-2} 5.3 · 10^{14} cm^{-2} 9.7 · 10^{14} cm^{-2}

Fig. 6.4 Pictures of a KU1-silica sample during irradiation with 4 MeV He$^+$ ions

intensity of each band as a function of the irradiation fluence) what we have studied in Chap. 7.

6.2 Ionoluminescence in Silica at Low Temperature

As it has been mentioned before, the origin of the band at 2.7 eV (~460 nm, the blue bland) is still a matter of conflict (see Sect. 1.2.1).

This band has been sometimes attributed to an intrinsic emission arising from a radiative transition of the STE, having a Full Width at Half Maximum (FWMH) of around 0.8–0.9 eV [6, 7], in accordance with low temperature irradiation experiments [8–10] and theoretical analyses [6, 10–12]. Some examples can be given here to illustrate the discrepancy existing about the position and origin of this band. Messina et al. [9] measured photoluminescence in silica observing a broad band at 2.5 eV that they associated to the STEs. Tanimura et al. [13] studied the luminescence induced by electron pulse irradiation in silica, they observed a peak located between 2.05 and 2.40 eV (the position of the peak depending on the delay time after the pulse); they ascribed this band to the STEs recombination. Also for quartz several studies using different techniques have reported an emission band at around 2.8 eV and have ascribed this band to the STEs: Hayes [14] used the Optically Detected Magnetic Resonance (ODMR) technique, Luff et al. [7] studied the cathodoluminescence, and Itoh et al. [8] irradiated with electron pulses and they already mention the conflict about the origin of this band. The band at 2.8 eV for quartz has also been predicted and related to STEs by theoretical studies, e.g., Van Ginhoven et al. [11] performed

Density Functional Theory (DFT) calculations and Ismail-Beigi et al. [10] used a many-electron Green's function approach.

However, other authors performing RT irradiations with light ions (mostly hydrogen and helium) assigned the 2.7 eV emission in silica with a FWMH of around 0.4 eV to a luminescence triplet-to-singlet transition of the oxygen-vacancy (ODC) centers [4, 15, 16]. The two emissions of the ODCII centers shown in Fig. 6.2 (one of them being at 2.7 eV) were also mentioned by Trukhin [5] and Skuja et al. [17].

To help clarifying this controversy, ionoluminescence experiments in silica at low temperature (100 K) have been performed in the framework of this thesis. These experiments were carried out at the CMAM implantation beamline (pp. 51 and 61). Two different types of silica (KU1 having a high OH content and IR with a low OH content) have been studied. The experimental conditions for these irradiations have been explained in p. 61. We measured the ionoluminescence at RT and at temperatures closed to liquid-nitrogen temperature (at 100 K, sometimes named LNT in this thesis for simplicity) with 1 MeV protons and with 40 MeV Br^{7+} ions under the same experimental conditions. Only the results for protons are shown here since the Br results did not provide any additional information for the problem we are dealing with here. Nevertheless, the results obtained with Br were in very good agreement with the proton results. The proton beam current was kept constant at around 110–115 nA. The irradiation flux was $\sim 4 \times 10^{11}$ $s^{-1}cm^{-2}$. The ionoluminescence was measured up to an irradiation fluence of $7 \times 10^{14} cm^{-2}$.

The spectra obtained for both samples at RT and LNT and at two different fluences (low and high) are shown in Fig. 6.5. The first clear effect that one can observe as a consequence of the temperature, is that at LNT the luminescence intensity is always higher than at RT (for both samples and during all the irradiation); non-radiative processes are more important at RT than at LNT, diminishing the IL intensity. At high fluences (Fig. 6.5b, d), apart from the different intensity, no special features are observed: for both temperatures the spectra consist of the same two peaks explained in Sect. 6.1. However, at low fluence (Fig. 6.5a, c), the behavior at RT and at LNT is very different: right at the beginning of the irradiation, a new emission band appears at around 2.2 eV for both samples. This new band disappears very quickly when the irradiation fluence increases.

To determine the features of the new emission observed in Fig. 6.5a, c, the LNT IL spectra have been fitted with different Gaussian curves using the program Fityk [2], as it had been done for the RT spectrum (p. 86). Two spectra of each sample (separated only by a few seconds during the irradiation) have been fitted in order to get a better determination of the position and the width of the peaks. The time after the beginning of the irradiation and corresponding fluences for each spectrum in Fig. 6.6 are summarized in the Table 6.3:

The four fitted spectra are shown in Fig. 6.6 with their decomposition in Gaussian curves. In this figure the green lines correspond to the experimental spectra, the black lines correspond to the fitted spectra obtained with Fityk, and the red peaks correspond to the Gaussian functions obtained for each emission band.

The spectra have been fitted with three Gaussian curves. The results of each fit in Fig. 6.6 are shown in the following Tables 6.4, 6.5, 6.6 and 6.7.

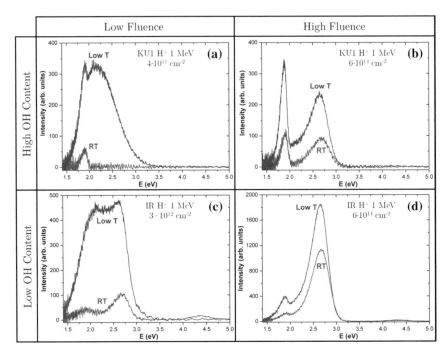

Fig. 6.5 Ionoluminescence spectra obtained at RT and at 100 K for **a** KU1 silica at low fluence, 4×10^{11} cm^{-2}; **b** KU1 silica at high fluence, 6×10^{14} cm^{-2}; **c** IR silica at low fluence, 3×10^{12} cm^{-2}; and **d** IR silica at high fluence, 6×10^{14} cm^{-2}

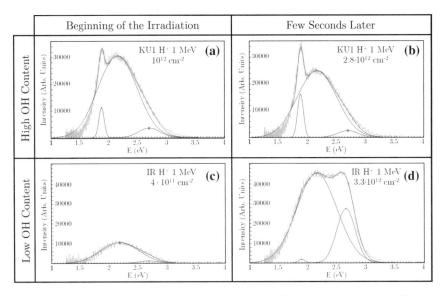

Fig. 6.6 Fits of the silica IL spectra at 100 K for **a** KU1 at 10^{12} cm^{-2}, **b** KU1 at 2.8×10^{12} cm^{-2}, **c** IR at 4×10^{11} cm^{-2}, and **d** IR at 3.3×10^{12} cm^{-2}

Table 6.4 Parameters of the IL peaks at 100 K obtained from the fitting of the KU1 spectrum presented in Fig. 6.6a

KU1 - 10^{12} cm^{-2}	Peak 1	Peak 2	Peak 3
Center (eV)	1.89	2.17	2.71
Center (nm)	656	571	458
Height (a.u.)	11211	30546	3588
HWHM (eV)	0.05	0.40	0.22

Table 6.5 Parameters of the IL peaks at 100 K obtained from the fitting of the KU1 spectrum presented in Fig. 6.6b

KU1 - $2.8 \cdot 10^{12}$ cm^{-2}	Peak 1	Peak 2	Peak 3
Center (eV)	1.88	2.17	2.70
Center (nm)	660	571	459
Height (a.u.)	16071	24918	2563
HWHM (eV)	0.06	0.40	0.21

Table 6.6 Parameters of the IL peaks at 100 K obtained from the fitting of the KU1 spectrum presented in Fig. 6.6c

IR - $4 \cdot 10^{11}$ cm^{-2}	Peak 1	Peak 2	Peak 3
Center (eV)	–	2.19	2.69
Center (nm)	–	566	461
Height (a.u.)	–	10491	1181
HWHM (eV)	–	0.39	0.20

Table 6.7 Parameters of the IL peaks at 100 K obtained from the fitting of the KU1 spectrum presented in Fig. 6.6d

IR - $3.3 \cdot 10^{12}$ cm^{-2}	Peak 1	Peak 2	Peak 3
Center (eV)	1.90	2.16	2.66
Center (nm)	653	574	466
Height (a.u.)	1854	45478	27484
HWHM (eV)	0.07	0.41	0.19

For both samples, the positions of the three peaks are ~1.9, ~2.17 and ~2.7 eV. The FWHM (FWHM = 2xHWHM) for the new band at 2.17 eV is ~0.8 eV, while for the blue band at 2.7 eV the FWHM is ~0.4 eV. These values obtained here are in very good agreement with the expected values for the STE emission and for the ODC emission.

The difference in the ionoluminescence emission at low and room temperature can be observed even with the naked eye: for both samples a yellowish or greenish color was observed right at the beginning of the LNT irradiation. The color turned very quickly red for the KU1, and blue for the IR silica, as observed at RT (Fig. 6.4, p. 88). Figure 6.7 shows the evolution of the color that is observed at 80 K, in the case of the KU1 sample the colors are compared to what had been observed at RT (for the IR samples the color was always blue at RT). In fact, the energy of 2.17 eV ($\lambda \simeq 571$ nm) corresponds to a yellow (almost green) emission in the visible light spectrum. This yellow emission becomes orange when it is combined with the red emission in the KU1 sample (Fig. 6.7a). But in the IR samples, where the blue emission is more

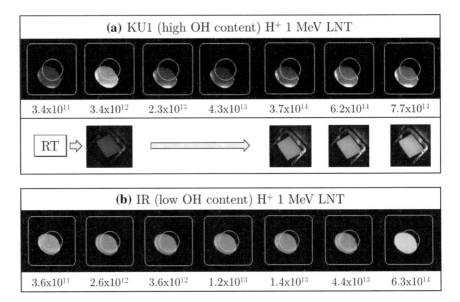

Fig. 6.7 Pictures of the silica samples during proton irradiation at 80 K. These pictures were taken with an optical camera placed at one of the viewports of the irradiation chamber. **a** KU1 sample, the images are compared here to the pictures obtained during RT irradiations at the same fluences. **b** IR sample (always blue at RT)

important than the red one, a greenish color is produced as a result of combining yellow and blue (Fig. 6.7b).

Figure 6.8 represents the intensity of the 2.17 eV band as a function of the irradiation fluence. This emission disappears at a fluence of $\sim 6.7 \times 10^{13}$ cm^{-2} for the KU1 sample and at a fluence of $\sim 4.5 \times 10^{13}$ cm^{-2} for the IR sample. After these fluences, the increase of the intensity is due to the contribution of the blue emission which becomes more important with fluence. For the 40 MeV Br irradiations mentioned before, the 2.17 eV (~ 571 nm) band was also observed at the beginning of the irradiation for both types of silica, and the emission also disappeared very quickly when increasing fluence.

These new results at low temperature confirm that the blue emission (2.7 eV) is due to the STE recombination at the ODCs, while the intrinsic recombination of STEs gives rise to a band at ~ 2.17 eV. The RT and low temperature (100 K) data obtained in this thesis clearly support that both, the red and blue emissions, arise from extrinsic STE recombination at irradiation-induced color centers, NBOH centers for the red one and ODCII for the blue one [3]. The physical model presented in this work involves the rapid self-trapping of the electron-hole (e-h) pairs created by irradiation, to form self-trapped excitons (STEs) [6, 7], their hoping migration through the silica network and their recombination at the suitable recombination centers.

At variance with the LNT experiments [18], our RT data support that extrinsic recombination at irradiation-induced color centers (NBOH and ODCII) [3] appears,

Fig. 6.8 Kinetics of the 2.17 eV emission (luminescence intensity vs fluence) for 1 MeV H⁺ irradiations at 100 K for **a** KU1 silica, and **b** IR silica

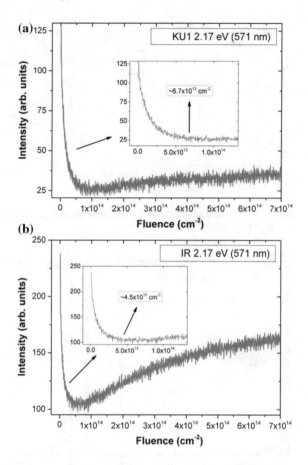

indeed, to be predominant in all irradiation cases. In fact, the absence of the emission due to the intrinsic recombination of the STEs in our experiments at RT is attributed to the very short STE decay time at or above RT [6, 7]. This decay time is ∼1 s at LNT, but only ∼1 ms at RT. When the number of point defects (such as NBOHCs and ODCs) produced by the ion impacts becomes important (i.e., fluences around 4×10^{13}–7×10^{13} cm^{-2} in the case of our irradiation conditions), the STEs quickly find point defects where they recombine, and thus, the 2.17 eV emission disappears.

6.3 Contribution of the Nuclear Stopping Power to the Ionoluminescence Signal

The IL results shown in this thesis were all measured with ions in the electronic stopping regime ($S_e \gg S_n$). A question that can arise when studying the results is

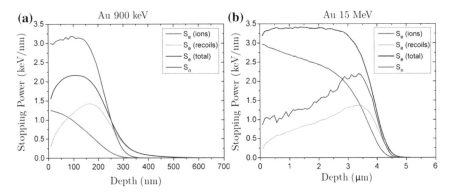

Fig. 6.9 Electronic and nuclear stopping powers obtained with SRIM [19, 20] for **a** 900 keV Au, and **b** 15 MeV Au in silica

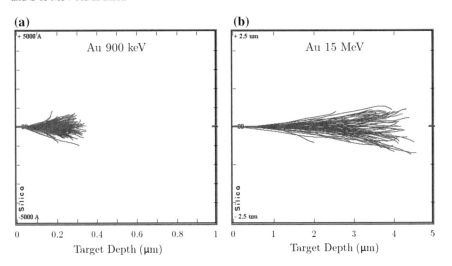

Fig. 6.10 Ion trajectories calculated with SRIM [19, 20] for **a** 900 keV Au, and **b** 15 MeV Au in silica

"how can the nuclear contribution be separated from the electronic one?" or "how can one be sure that the nuclear contribution does not affect the IL spectra?". Some IL experiments are presented here to prove that the nuclear contribution to the IL spectra is negligible compared to the electronic excitation one.

The experiments were performed at the IBML in Knoxville (Sect. 4.3, p. 54). We measured the ionoluminescence of silica produced with the same element (Au) but with two different energies: 900 keV and 15 MeV. Figures 6.9 and 6.10 show the stopping powers and the ion trajectories, respectively, for both cases (900 keV Au$^+$, and 15 MeV Au^{5+}). For 900 keV Au the nuclear contribution (S_n) is predominant, while for 15 MeV Au the electronic stopping power (S_e) dominates. However, the other contributions are also present in both cases, but the predominance

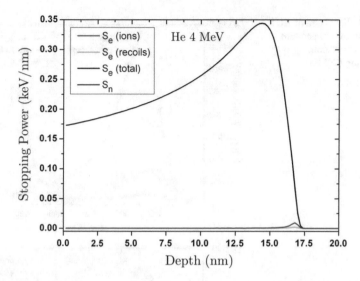

Fig. 6.11 Electronic and nuclear stopping powers obtained with SRIM [19, 20] for 4 MeV He in silica. S_e from the ions practically coincides with the total S_e

of the nuclear or the electronic regime is clear in each case. Due to the features of the accelerator, 900 keV and 15 MeV was the higher difference on the energy that we could get for the same ion.

The ionoluminescence of silica was also measured for 4 MeV He$^+$ in order to compare the results and the scale of the spectra to the other results obtained in this thesis (Chaps. 7 and 8). In the case of 4 MeV He, the nuclear stopping power is completely negligible as it can be seen in Fig. 6.11, but its mass is very different from the Au mass, and the idea here is to fix the maximum number of parameters in order to compare only the effect of the nuclear end electronic contributions.

The experiments were carried out under the same irradiation conditions (beam size, geometry, spectrophotometer, etc.) at two different temperatures: RT ($\sim$295 K) and low temperature ($\sim$133 K). The integration time was 500 ms for the He and the 900 keV Au, and 100 ms for the 15 MeV Au (the IL intensity for 15 MeV Au has been always multiplied by 5 to get a correct comparison). The irradiation area was always 3×3 mm^2. The ion fluxes were 7×10^{11} cm^{-2}s^{-1} for He ($I = 10$ nA), 1.4×10^{12} cm^{-2}s^{-1} for 900 keV Au ($I = 20$ nA), and 2.8×10^{11} cm^{-2}s^{-1} for 15 MeV Au ($I = 20$ nA). Although two types of silica were studied (KU1 and IR), only the IL spectra obtained for the KU1 samples are presented here to simplify the understanding of the results.

Figure 6.12 shows the IL spectra (as a function of λ) at relatively high fluence (4.6×10^{14} cm^{-2}) at low temperature and RT. In all the spectra the two bands at 460 and 650 nm (2.7 and 1.9 eV) are observed. The general features are the same for both temperatures: the IL intensity is higher for 15 MeV Au than for 900 keV Au.

Fig. 6.12 IL spectra of KU1
silica at **a** low temperature
(133 K), and **b** room
temperature (295 K)

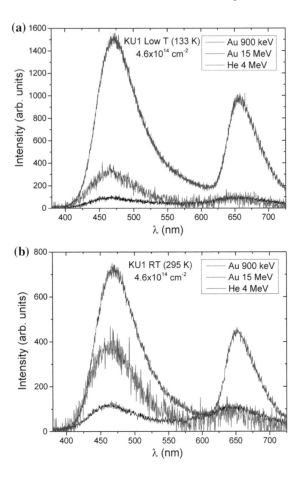

For a better comparison of the IL intensity during all the irradiation time, the
kinetics (intensity vs fluence) of the main band (2.7 eV) for the three ions and for both
temperatures are represented in Fig. 6.13. For 15 MeV Au the maximum irradiation
fluence was $10^{14}\,\mathrm{cm}^{-2}$, but after this fluence the level of the IL intensity remains
constant.

The features of the kinetics curves for the different ions will not be discussed
here since the role of the ion mass and the electronic stopping power will be studied
in detail in Chap. 7. The important fact that has to be appreciated from Fig. 6.13 is
that the IL intensity for the 900 keV irradiation is much lower than the intensity for
15 MeV irradiation. The intensity for 900 keV is ~30% of the intensity for 15 MeV
at both temperatures. Of course we still see an IL signal at 900 keV because there is
also an electronic contribution, but even in this case where the difference between S_e
and S_n is not so high, we can see the big difference in the IL intensity. In the light of
these results, we can now affirm that for our irradiations where $S_e \gg S_n$ (Chaps. 7
and 8) the nuclear contribution to the IL signal is negligible.

Fig. 6.13 IL intensity of the 2.7 eV band versus fluence for KU1 silica at **a** low temperature (133 K), and **b** room temperature (295 K). The beginning of the peak cannot be observed with 900 keV Au because the integration time was 500 ms, while it was only 100 ms for 15 MeV Au

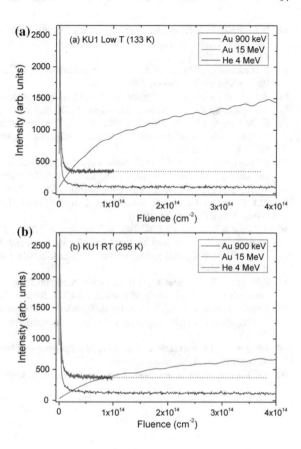

References

1. Y. Wang, P.D. Townsend, Common mistakes in luminescence analysis. J. Phys. Conf. Ser. **398**(1), 012003 (2012)
2. Fityk, http://fityk.nieto.pl/
3. L. Skuja, M. Hirano, H. Hosono, K. Kajihara, Defects in oxide glasses. Physica Status Solidi (c) **2**(1), 15–24 (2005)
4. S. Nagata, S. Yamamoto, A. Inouye, B. Tsuchiya, K. Toh, T. Shikama, Luminescence characteristics and defect formation in silica glasses under H and HE ion irradiation. J. Nucl. Mater. **367–370**(B), 1009–1013 (2007). Proceedings of the Twelfth International Conference on Fusion Reactor Materials (ICFRM-12)
5. A.N. Trukhin, Luminescence of localized states in silicon dioxide glass. A short review. J. Non-Cryst. Solids **357**(8–9), 1931–1940 (2011). SiO2, Advanced Dielectrics and Related Devices
6. A.K.S. Song, R.T. Williams, *Self-Trapped Excitons* (Springer, Berlin, 1996)
7. B.J. Luff, P.D. Townsend, Cathodoluminescence of synthetic quartz. J. Phys. Condens. Matter **2**, 8089–8097 (1990)
8. C. Itoh, K. Tanimura, N. Itoh, Optical studies of self-trapped excitons in SiO2. J. Phys. C Solid State Phys. **21**(26), 4693–4702 (1988)

9. F. Messina, L. Vaccaro, M. Cannas, Generation and excitation of point defects in silica by synchrotron radiation above the absorption edge. Phys. Rev. B Condens. Matter Mater. Phys. **81**, 035212 (2010)

10. S. Ismail-Beigi, S.G. Louie, Self-trapped excitons in silicon dioxide: mechanism and properties. Phys. Rev. Lett. **95**, 156401 (2005)

11. R.M. Van Ginhoven, H. Jónsson, L. René Corrales, Characterization of exciton self-trapping in amorphous silica. J. Non-Cryst. Solids **352**(23–25), 2589–2595 (2006)

12. K. Awazu, S. Ishii, K. Shima, S. Roorda, J.L. Brebner, Structure of latent tracks created by swift heavy-ion bombardment of amorphous sio_2. Phys. Rev. B Condens. Matter Mater. Phys. **62**, 3689–3698 (2000)

13. K. Tanimura, C. Itoh, N. Itoh, Transient optical absorption and luminescence induced by band-to-band excitation in amorphous SiO_2. J. Phys. C Solid State Phys. **21**(9), 1869–1876 (1988)

14. W. Hayes, Mechanisms of exciton trapping in oxides. J. Lumin. **31–32**, Part 1:99–101 (1984)

15. P.D. Townsend, P.J. Chandler, L. Zhang, *Optical Effects of Ion Implantation* (Cambridge University Press, Cambridge, 1994)

16. S. Nagata, S. Yamamoto, K. Toh, B. Tsuchiya, N. Ohtsu, T. Shikama, H. Naramoto, Luminescence in SiO_2 induced by MeV energy proton irradiation. J. Nucl. Mater. **329–333**(B), 1507–1510 (2004). Proceedings of the 11th International Conference on Fusion Reactor Materials (ICFRM-11)

17. L.N. Skuja, A.N. Streletsky, A.B. Pakovich, A new intrinsic defect in amorphous SiO_2: twofold coordinated silicon. Solid State Commun. **50**(12), 1069–1072 (1984)

18. J.M. Costantini, F. Brisard, G. Biotteau, E. Balanzat, B. Gervais, Self-trapped exciton luminescence induced in alpha quartz by swift heavy ion irradiations. J. Appl. Phys. **88**, 1339–1345 (2000)

19. J.F. Ziegler, J.P. Biersack, U. Littmark, *The Stopping and Range of Ions in Solids* (Pergamon, New York, 1985), http://www.srim.org

20. J.F. Ziegler, SRIM: The Stopping and Range of Ions in Matter, http://www.srim.org/

Chapter 7
Ionoluminescence in Silica: Role of the Silanol Group Content and the Ion Stopping Power

The purpose of this chapter is to report on a comparison of the IL data obtained under light and heavy ion irradiations as a means to reveal the different physical processes operating in each case. In all cases, recombination of self-trapped excitons (STEs) at color centers, Non-Bridging Oxygen Hole Centers (NBOHCs) and Oxygen-Deficient Centers (in particular, ODCII), created by irradiation, are assumed to be the predominant light emission process for the red (1.9 eV) and blue (2.7 eV) emissions, respectively. However, the comparison of the IL kinetics under light and heavy ion irradiations shows remarkable differential features. In particular, it reveals a coupling between the irradiation-induced structural damage caused by swift-heavy ion (SHI) irradiation and light emission. This coupling is absent for light ions due to their much lower structural disorder on the SiO_2 network. Detailed spectroscopic experiments have shown, indeed, significant distortions and changes in the ring-size distribution during SHI irradiation measured through the frequency of the ω_4 mode [1] or by Raman spectroscopy [2]. In our experiments we have observed a definite correlation between those changes in ω_4 and the shape of the IL kinetic curves, suggesting that the network distortions modify the migration of the STEs to the recombination centers, either NBOHCs or ODCIIs. On the other hand, in order to separate the effect of pre-existing defects on the IL results, comparative irradiation experiments have been performed on samples containing different amounts of OH groups that constitute a very common manufacturing product in silica. It is well known that silanol groups $(Si–OH)$ determine the optical properties [3] of silica and so are very relevant to technological applications. The results obtained in this chapter remark the important role of these silanol groups on the light emission yields.

The irradiation experiments were carried out on three different types of silica (KU1, UV, IR) and their main difference is the amount of OH groups that they contain. Table 7.1 summarizes the three types of silica and their OH content estimated from their IR absorption spectra as explained in Sect. 2.1.3 (p. 21).

The measurements reported in this chapter were carried out at RT at the standard chamber at CMAM (pp. 50 and 61).

© Springer Nature Switzerland AG 2018
D. Bachiller Perea, *Ion-Irradiation-Induced Damage in Nuclear Materials*,
Springer Theses, https://doi.org/10.1007/978-3-030-00407-1_7

Table 7.1 List of the three types of silica used and their OH content experimentally estimated from their IR absorption spectra

Type of silica	OH content (ppm)
KU1	$(1.34 \pm 0.10) \times 10^3$
UV crystran	(573 ± 42)
IR crystran	(13 ± 1)

Table 7.2 Stopping powers and ion ranges of the ions used in this work, calculated with SRIM [4, 5]

Ion	E (MeV)	S_e (keV/nm)	S_n (keV/nm)	S_T (keV/nm)	Ion range (μm)
H^+	2.0	0.027	2×10^{-5}	0.027	46
He^+	4.0	0.173	10^{-4}	0.173	16.7
C^+	4.0	1.291	0.003	1.294	4.2
Si^{6+}	28.6	3.417	0.006	3.423	10.2
Br^{4+}	18.0	4.959	0.104	5.063	6.3
Br^{6+}	28.0	6.066	0.073	6.139	8.1

Irradiations were performed with several ions and energies covering the range from H^+ at 2 MeV to Br^{6+} at 28 MeV and currents below 40 pnA (particle nanoampere) (ion fluxes around 10^{12} cm^{-2}s^{-1}) to avoid overheating of the samples (except in the case of protons at high fluence, where the current was 400 nA). The corresponding electronic (S_e) and nuclear (S_n) stopping powers at the input face are listed in Table 7.2, together with the ion ranges. It can be seen there that the electronic stopping power S_e is clearly dominant in all cases as it is also illustrated with the three examples in Fig. 7.1.

7.1 IL Spectra

Figure 7.2 shows a simple example of the IBIL spectra obtained with 2 MeV protons for the three types of silica at a fluence of 3×10^{14} cm^{-2}. Although the position of the main peaks are roughly the same, their intensity is completely different. The red band (1.9 eV) is predominant in the sample with a high OH content (KU1, pink line) while the blue band (2.7 eV) is much more intense in the IR sample with a low OH content (green line). The UV sample shows an intermediate behavior since it contains a medium level of OH impurities.

In order to illustrate the relevant role of the OH contents and ion mass on the IL spectra we show in Fig. 7.3 emission spectra (as a function of the wavelength) corresponding to RT irradiations with light (H) and heavy (Br) ions for the three types of samples (KU1, UV Crystran, and IR Crystran). The spectra cover the range from 200

Fig. 7.1 Electronic and
nuclear stopping powers of
2 MeV protons, 4 MeV He$^+$
and 18 MeV Br^{3+} in
amorphous silica

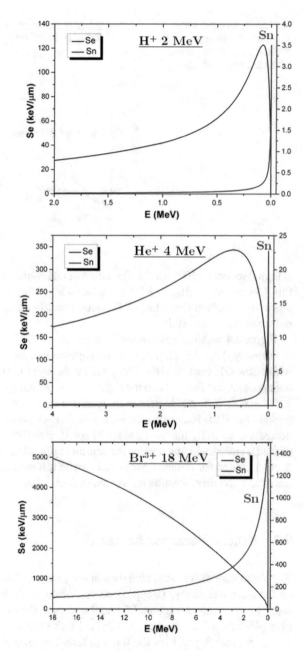

to 900 nm. We compare for each case a spectrum at low fluence ($\sim 10^{12}$–10^{13} cm^{-2})
to a spectrum at high fluence ($\sim 10^{14}$ cm^{-2}). The blue band has its maximum at
460 nm for protons, but it is slightly displaced to lower wavelengths for Br, reaching
its maximum at 450 nm for low fluences ($\sim 10^{12}$–10^{13} cm^{-2}) and moving to 455 nm

Fig. 7.2 IL spectra (as a
function of the energy)
obtained with 2 MeV protons
for the three types of silica

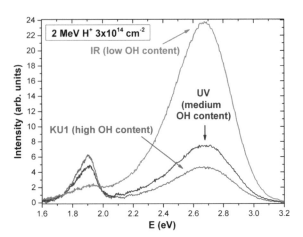

at high fluences ($\sim 10^{14}$ cm^{-2}). The effect of OH contents is illustrated by comparing
Fig. 7.3a corresponding to KU1 samples to Fig. 7.3b, c (UV and IR Crystran samples). We conclude that a high OH content clearly enhances the contribution of the
red versus the blue band.

Figure 7.4 summarizes some of the spectra of Fig. 7.3 (in this case as a function
of the energy) for the samples with the highest OH content (KU1) and for the sample
with a low OH content (IR). The plots on the left (a, c) show the spectra obtained
with protons and the plots on the right (b, d) show the spectra obtained with Br ions.
Two fluences (low and high) are compared in each plot. It can be observed that with
protons the IBIL intensity increases with fluence while with Br ions the intensity
decreases. To study the dependence of the IL intensity on the fluence (the kinetics
of the IL emissions), the area of the central region of each peak (1.9 and 2.7 eV) has
been measured for different irradiations (with different ions and energies) during all
the irradiation time. Results are shown in Sect. 7.2.

7.2 Kinetic Behavior for the IL

The significant differences observed in the kinetics of the IL emissions, depending
on the mass and energy (stopping power) of the projectile ion, constitute the main
focus of this chapter. Figures 7.5 and 7.6 display the evolution with fluence for the
blue (450 nm, 2.7 eV) and red (650 nm, 1.9 eV) band heights for different ions (H,
He, C, Si, and Br) and for the three different types of silica studied in this thesis.
One first observes that the situation is complex and that the two effects (OH content
and projectile ion stopping power) are heavily intermixed. The main results to be
remarked are described next.

For heavy mass ion irradiations there is an initial rapid growth of the two yields
with fluence that reach a maximum at a fluence Φ_{Max} (around 10^{12}–10^{13} cm^{-2}

Fig. 7.3 Emission spectra (as a function of the wavelength) obtained under irradiation with 2 MeV H⁺ and 24 MeV Br⁵⁺ beams at low ($\sim 10^{12}$–10^{13} cm⁻²) and high ($\sim 10^{14}$ cm⁻²) fluence. Note the different scale used in (c) for protons

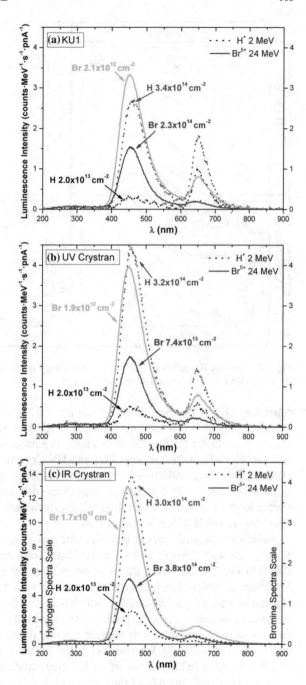

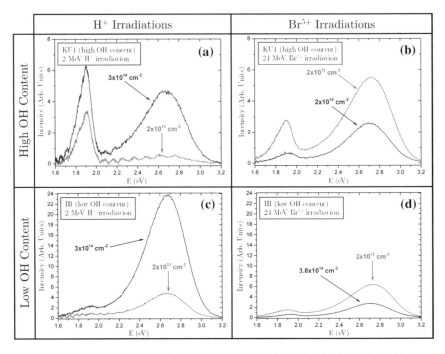

Fig. 7.4 Emission IL spectra as a function of the energy obtained under irradiation with 2 MeV H$^+$ and 24 MeV Br^{5+} beams at low and high fluence

depending on the stopping power of the particle) followed by a slower decrease in yield, as reported in a previous work [6]. In the case of He irradiations we do not see a maximum for the blue band, but there is a maximum in the red band for the samples with a high OH content (KU1 and UV Crystran). Under H irradiation, the yields of the two IL emissions increase monotonically with fluence in the range up to 3.6 × 10^{14} cm^{-2} with a decreasing growth rate that may suggest a final saturation level at higher fluences. The maximum intensity of both bands is always higher for light ions (H and He) than for heavy ions.

To clarify the high fluence IL behavior with protons we have performed irradiation experiments with fluences up to 6 × 10^{16} cm^{-2}, and the corresponding results are shown in Fig. 7.7. We found the same behavior as for He irradiations: there is no maximum in the blue band, and there is a maximum in the red band only in the case of samples with a high OH content. Both bands start from a very low yield, but the red band experiences a faster growth at low fluence. As we will discuss in Sect. 7.3 the maximum of the intensity is reached at lower fluences for the red band than for the blue band (Table 7.3). We also found that the red band reach a saturation level which is the same independently of the OH content of the sample (Fig. 7.7b).

In what concerns the role of the OH contents, it is worth noting that both bands show a similar trend for the three types of samples, except in the case of the red band when irradiating with light ions (He and H). However, we can appreciate significant

Fig. 7.5 Evolution of the blue band height (2.7 eV) with fluence for different irradiations and samples

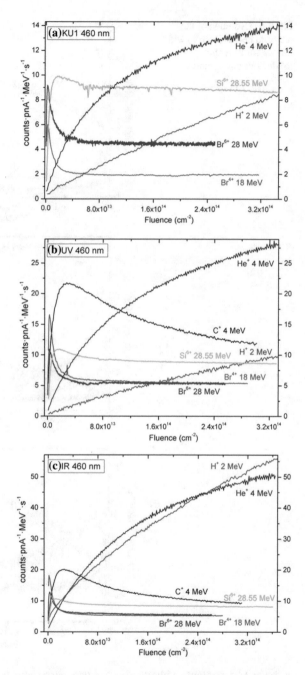

quantitative differences in the intensity of the bands. One notes that the ratio between the emission yields of the red and blue band is strongly enhanced for high OH contents. In other words, the blue band reaches higher values for the samples having

Fig. 7.6 Evolution of the red band height (1.9 eV) with fluence for different irradiations and samples

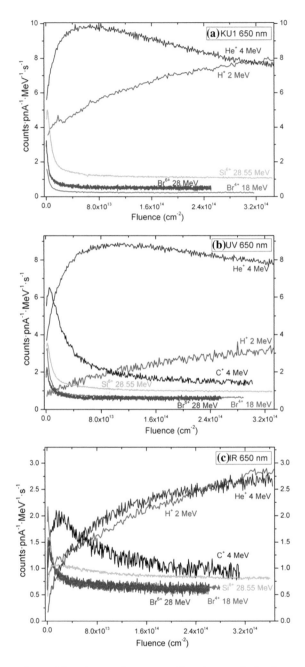

a lower OH content, whereas the red band experiences a rather inverse behavior: its intensity increases when increasing the OH content. This behavior is indeed consistent with the data shown in Table 7.1.

Fig. 7.7 Evolution of the height of the red and blue IL bands for proton irradiations in the high fluence region up to $6 \times 10^{16}\,\mathrm{cm}^{-2}$

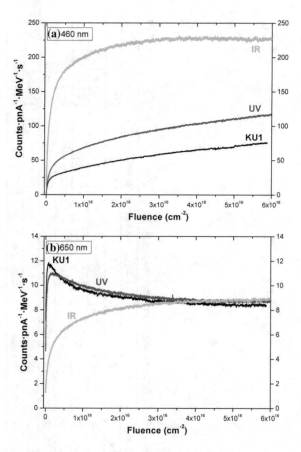

An interesting feature, not previously reported and somewhat hidden in our spectra, is an initial jump in the red emission yield that apparently corresponds to a very fast process. It occurs for samples with high OH contents and essentially disappears for OH-free samples. This behavior is only observed for bromine ions in the case of the blue band, and even in this case, the maximum is produced at higher fluences (Φ_{Max}) than for the red band (Table 7.3). In principle, one may suggest that the silanol $Si{-}OH$ groups provide an additional channel for the generation of NBOH centers under irradiation and, thus, an enhanced red emission.

7.3   Dependence of the Maximum Intensity with the Stopping Power

As seen in Figs. 7.5, 7.6 and 7.7 the fluence at which the maximum intensity is produced (Φ_{Max}) decreases strongly with increasing mass and energy of the projectile ion, i.e., with increasing electronic stopping power. The values of Φ_{Max} are listed

Table 7.3 Values of the electronic stopping power (S_e) and Φ_{Max} for the different irradiations and samples [4, 5]

Ion and energy	S_e (keV/nm)	Φ_{Max} (ion/cm^2) (460 nm)			Φ_{Max} (ion/cm^2) (650 nm)		
		KU1	UV	IR	KU1	UV	IR
Br^{6+} 28 MeV	6.068	2.3×10^{12}	2.3×10^{12}	2.5×10^{12}	5.0×10^{11}	1.3×10^{12}	1.6×10^{12}
Br^{4+} 18 MeV	4.959	2.4×10^{12}	2.4×10^{12}	1.2×10^{12}	$<5.0 \times 10^{11}$	1.7×10^{12}	1.2×10^{12}
Si^{6+} 28.55 MeV	3.417	1.8×10^{13}	1.5×10^{13}	1.2×10^{13}	2.0×10^{12}	2.0×10^{12}	1.4×10^{13}
C$^+$ 4 MeV	1.291	–	2.8×10^{13}	2.5×10^{13}	–	4.8×10^{12}	2.0×10^{13}
He$^+$ 4 MeV	0.173	No max	No max	No max	6.5×10^{13}	1.0×10^{14}	No max
H$^+$ 2 MeV	0.027	No max	No max	No max	8.1×10^{14}	1.7×10^{15}	No max

Fig. 7.8 Φ_{Max} versus the ion stopping power for the peak at 460 nm (blue peak) compared to Awazu's data. Note that some points overlap due to the scale so we cannot see clearly the black square at S = 3.4 keV/nm because it is behind the red circle

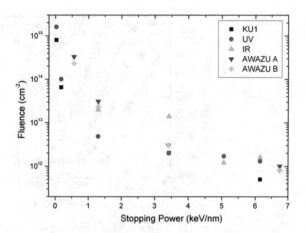

in Table 7.3, and the data for the blue emission are plotted in Fig. 7.8. A semilog fluence scale has been used to plot the data since the change in Φ_{Max} spans over several orders of magnitude. The plot of the data for the red emission is not shown here because its behavior is very similar to the blue emission's one and it does not provide any different information. The value of Φ_{Max} is not very different for the blue and red bands although it appears slightly smaller for the red emission.

In Fig. 7.8 we have compared our results with the fluences at which the frequency of the ω_4 band changes abruptly its decreasing rate (Fig. 7.9, obtained from [1]). We have calculated these values from Fig. 7.9 in two different ways: fluence at which the change in the slope is produced (AWAZU A) and fluence when the prolongations of the sharp decreasing curves intersect the maximum value (1078 cm^{-1}) (AWAZU B). We can observe that both values are very similar. Our results are therefore in very good agreement with those obtained by Awazu et al. [1], we observe the same dependence on stopping power of the fluence at which changes are observed in both cases.

7.4 Discussion

7.4.1 Role of the OH Content

The experimental data show that the kinetic behaviors of the IL emissions present specific features in samples with high OH contents. In fact, from the values of Φ_{Max} in Table 7.3 we observe that, for samples with high OH content, the curves for the red emission present a faster initial growth than the blue emissions (Φ_{Max} is lower for 650 nm than for 460 nm). This difference in the Φ_{Max} value for the red and blue emissions is not found for the samples containing a low OH concentration. If we compare an irradiation in particular, we can see that Φ_{Max} is also lower for the

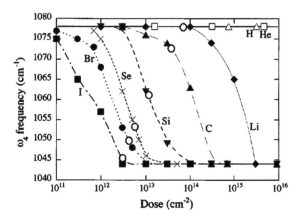

Fig. 7.9 Figure from [1]. Original caption: Frequency at an absorption maximum of the ω_4 band against dose. 10 MeV H open squares; 15 MeV He, open triangles; 2 MeV Li, closed diamons; 4 MeV C, closed triangles; 30 MeV Si, closed and reversed triangles; 35 MeV Se, crosses; 67 MeV Br, closed circles; 80 MeV I, closed squares. All lines are a guide for the eyes. Open circles correspond to the bond energy of $Si-O$ which was obtained from the electronic stopping power of each ions multiplied by dose

red band in samples with a high OH content that in IR samples, while it is of the same order of magnitude in the case of the blue band. As anticipated in Sect. 7.2, this behavior, not reported so far, could be expected if one considers an efficient additional channel for generation of NBOH centers due to the scission of silanol groups ($Si-OH$) [7, 8]:

$$\equiv Si - O - H \Rightarrow \ \equiv Si - O - O^{\,\bullet} + H \qquad (7.1)$$

The initial abrupt jump in the yield for the red luminescence could be associated to the radiative decay of the NBOH centers generated by process (7.1) in an excited state. Note that this process cannot be detected in standard optical absorption measurements performed just after irradiation due to the rapid thermal recovery of the silanol groups at RT [9]. This extrinsic formation channel adds to the well-established intrinsic channels involving the breaking of, possibly strained, $Si-O-Si$ bonds [10, 11] operating in the three types of samples:

$$\equiv Si - O - Si \equiv \ \Rightarrow \ \equiv Si - O^{\,\bullet} \,(NBOH) + \,^{\bullet}Si \equiv (E') \qquad (7.2)$$

$$\equiv Si - O - Si \equiv \ \Rightarrow \ \equiv Si - Si \equiv (ODCI) + O\,(int) \Rightarrow 2 = Si^{\,\bullet\,\bullet} \,(ODCII) + O\,(int) \qquad (7.3)$$

where int stands for interstitial. Process (7.2) leads to the creation of a coupled pair of E' ($\equiv Si^{\,\bullet}$) and NBOH centers ($\equiv Si-O^{\,\bullet}$), whereas channel (7.3) is responsible for the creation of ODCII centers ($= Si^{\,\bullet\,\bullet}$) and interstitial oxygen atoms. If the sample does not contain OH impurities, there is just a competition between the two above channels (7.2) and (7.3), but channel (7.1), enhances the red emission, leading to a different kinetic behavior of the red and blue emissions. In fact, the operation of the extrinsic channel (7.1) drives the STEs into the $Si-OH$ sites, and reduces their trapping at the strained intrinsic sites required for the operation of channels (7.2) and (7.3). An important conclusion of this result is that the IL data support an electronic mechanism for color center formation induced by the localization of STEs at either silanol bonds [7, 8] or strained intrinsic bonds [12, 13]. The excitation and subsequent de-excitation of these bonds give rise to light emission (radiative process) and defect formation (non-radiative process). Therefore, the observed behavior in our experiments supports an excitonic mechanism and it does not appear consistent with defect formation by elastic collision processes and even with thermally-induced scission of intrinsic bonds. Obviously, color centers created by collisions should contribute to the overall damage and may be responsible for the high rates of defect formation measured in the experiments with protons at high fluences (up to 6 × 10^{16} cm^{-2}).

7.4.2 Role of the Electronic Stopping Power

In order to discuss the effect of the electronic stopping power a rather detailed analysis of the IL processes is necessary. Although this is out of the scope of this phenomeno-logical chapter, several qualitative considerations can be advance here. The three key processes to describe the IL output are the nature of the electronic excited states, their migration behavior through the material and the effective recombination centers for light emission. Our proposal is to consider that self-trapped excitons (STEs) form readily from the excited electron-hole pairs in SiO_2 and are the responsible agents for the transport of the excitation energy [12]. In accordance with the available evidence [14, 15], the STEs move by thermally-activated hoping through the SiO_2 network and should recombine, either at NBOH centers (red emission) or ODCII centers (blue emission) created by the irradiation. Therefore, the whole process should be very sensitive to the structural disorder created by the irradiation; this structural disorder has been investigated by spectroscopic techniques (IR absorption and Raman) and finally leads to the compaction of the SiO_2 network. In fact, detailed data [1] have revealed that the infrared frequencies for the ω_4 mode, associated to the Si–O bonds, strongly change under SHI irradiation but not for light ion irradiation. Moreover, the structural change starts at decreasing fluences for increasing ion stopping power. This behavior resembles that one found in our IL experiments showing that the maximum fluence at which the slope of the IL versus fluence is turned from positive to negative occurs at higher fluences the lower the stopping power is. In fact, the data by Awazu et al. [1] (Fig. 7.9) have been compared to the results shown in Fig. 7.5 showing a reasonable accordance with our data (Fig. 7.8). Those data have been determined from Fig. 5 of Ref. [1] through two different procedures: (A) fluence at which the change in the slope is produced and (B) fluence at which the prolongations of the sharp decreasing curves intersect the maximum value ($1078\,cm^{-1}$). Both values are very similar. Although additional work is necessary, our view is that the structural disorder caused by the SHI irradiations slow down the hoping migration of the STEs preventing that they reach the appropriate recombination centers and give rise to light emission. A deeper analysis of the processes described here will be developed in Chap. 8. Anyhow, the results shown in this chapter prove that IL is a promising approach to investigate in a quantitative way the generation of structural damage by SHI irradiation and its role on the electronic (excitonic) effects leading to defect formation and light emission.

References

1. K. Awazu, S. Ishii, K. Shima, S. Roorda, J.L. Brebner, Structure of latent tracks created by swift heavy-ion bombardment of amorphous SiO_2. Phys. Rev. B Condens. Matter Mater. Phys. **62**, 3689–3698 (2000)
2. R. Saavedra, M. León, P. Martín, D. Jiménez-Rey, R. Vila, S. Girard, A. Boukenter, Y. Querdane, Raman measurements in silica glasses irradiated with energetic ions. AIP Conf. Proc. **1624**, 118–124 (2014)
3. E. Vella, R. Boscaino, Structural disorder and silanol groups content in amorphous SiO_2. Phys. Rev. B Condens. Matter Mater. Phys. **79**(8), 085204 (2009)
4. J.F. Ziegler, J.P. Biersack, U. Littmark, *The Stopping and Range of Ions in Solids* (Pergamon, New York, 1985), http://www.srim.org
5. J.F. Ziegler, *SRIM: The Stopping and Range of Ions in Matter*, http://www.srim.org/
6. R.C. Ropp, *Luminescence and the Solid State* (Elsevier, Amsterdam, 2004)
7. K. Kajihara, L. Skuja, M. Hirano, H. Hosono, Formation and decay of nonbridging oxygen hole centers in SiO_2 glasses induced by F_2 laser irradiation: in situ observation using a pump and probe technique. Appl. Phys. Lett. **79**(12), 1757–1759 (2001)
8. K. Kajihara, Y. Ikuta, M. Hirano, T. Ichimura, H. Hosono, Interaction of F_2 excimer laser pulses with hydroxy groups in SiO_2 glass: hydrogen bond formation and bleaching of vacuum ultraviolet absorption edge. J. Chem. Phys. **115**(20), 9473–9476 (2001)
9. H. Hosono, K. Kajihara, T. Suzuki, Y. Ikuta, L. Skuja, M. Hirano, Vacuum ultraviolet optical absorption band of non-bridging oxygen hole centers in SiO_2 glass. Solid State Commun. **122**(3–4), 117–120 (2002)
10. T.E. Tsai, D.L. Griscom, Experimental evidence for excitonic mechanism of defect generation in high-purity silica. Phys. Rev. Lett. **67**, 2517–2520 (1991)
11. H. Hosono, Y. Ikuta, T. Kinoshita, K. Kajihara, M. Hirano, Physical disorder and optical properties in the vacuum ultraviolet region of amorphous SiO_2. Phys. Rev. Lett. **87**, 175501 (2001)
12. F. Messina, L. Vaccaro, M. Cannas, Generation and excitation of point defects in silica by synchrotron radiation above the absorption edge. Phys. Rev. B Condens. Matter Mater. Phys. **81**, 035212 (2010)
13. A.K.S. Song, R.T. Williams, *Self-Trapped Excitons* (Springer, Berlin, 1996)
14. N. Itoh, A.M. Stoneham, *Materials Modification by Electronic Excitation* (Cambridge University Press, Cambridge, 2001)
15. J.M. Costantini, F. Brisard, G. Biotteau, E. Balanzat, B. Gervais, Self-trapped exciton luminescence induced in alpha quartz by swift heavy ion irradiations. J. Appl. Phys. **88**, 1339–1345 (2000)

Chapter 8
Exciton Mechanisms and Modeling of the Ionoluminescence in Silica

In this chapter a theoretical model is presented to discuss detailed kinetic data describing the evolution of the two main ionoluminescence (IL) bands at 650 (1.9) and 460 nm (2.7 eV) in silica as a function of the irradiation fluence at room temperature.

Some of the data shown in Chap. 7 will be used in this chapter to introduce the physical model here presented. In particular, the results obtained for dry silica (IR, low OH content) will be evoked here as an introduction for the model, since this is the type of silica (of the three that were used for Chap. 7) with the highest purity.

A physical model is proposed to explain the kinetics of the main IL emissions of silica, and a preliminary mathematical formulation is deduced and proved by fitting some of the data obtained in Chap. 7.

8.1 Kinetic Behavior for the IL: Correlation with Structural (Macroscopic) Damage

This section summarizes the results obtained for IR silica in Chap. 7 in order to use them as a starting point for the theoretical model described in Sect. 8.2.

IL spectra of IR-silica samples under H (2 MeV) and Br (24 MeV) irradiation at RT are displayed in Fig. 8.1a and b respectively. The two projectile ions in Fig. 8.1 are representative examples of low and heavy mass ions, having electronic stopping powers of 0.03 and 5.7 keV/nm, respectively. The spectra show the two main bands explained in Sect. 6.1, as well as the small band at around 4.4 eV.

As it has been seen, significant differences are observed in the IL kinetics at RT, depending on the ion mass and energy (stopping power). Figure 8.2 shows the evolution with fluence for the heights of the two dominant bands under 2 MeV H (Fig. 8.2a) and 24 MeV Br (Fig. 8.2b) irradiations at RT for fluences up to 4×10^{14} cm^{-2}.

In all cases the initial emission yields of the two bands for a pristine sample is zero within the sensitivity of our set-up, pointing out to an extrinsic character

© Springer Nature Switzerland AG 2018
D. Bachiller Perea, *Ion-Irradiation-Induced Damage in Nuclear Materials*,
Springer Theses, https://doi.org/10.1007/978-3-030-00407-1_8

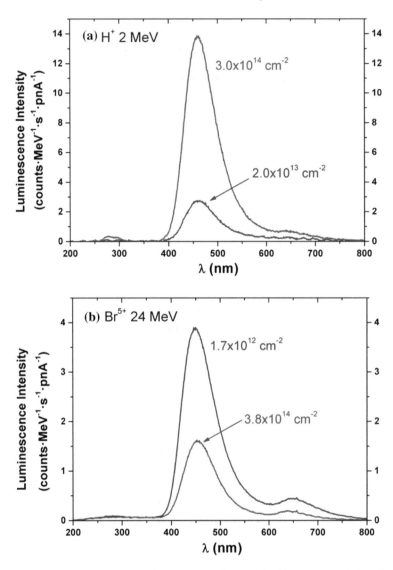

Fig. 8.1 Emission spectra at RT for silica samples irradiated with **a** H at 2 MeV and **b** Br at 24 MeV at two different fluences. Spectra show the two main bands at 2.7 eV (460 nm) and 1.9 eV (650 nm)

of the emissions. Under light ion (H) irradiation the intensity of the bands increases monotonically with fluence although a final saturation level is not observed at fluences below 5×10^{14} cm^{-2}. The growth rate is clearly smaller for the red than for the blue band. Additional data for protons at higher fluences (up to 6×10^{16} cm^{-2}) shown in Fig. 8.3, essentially confirm the occurrence of a saturation level for the two emissions, although at much higher values than reached in Fig. 8.2a. The data, all together, are in essential accordance with those reported by Nagata et al. [1], using a smaller fluence

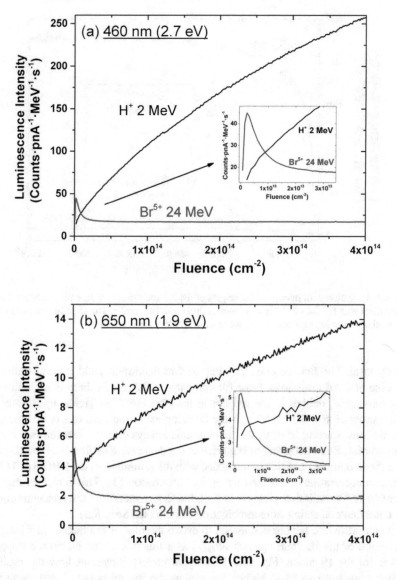

Fig. 8.2 Evolution with fluence of the heights of **a** the 2.7 eV and **b** the 1.9 eV IL emission bands for RT irradiations with H^+ ions at 2 MeV and Br^{5+} ions at 24 MeV. The insets show in more detail the kinetics observed at low fluence ($<5 \times 10^{13}$ cm^{-2})

range (up to around 10^{15} cm^{-2}). In this range they observed a saturation level for the red band but not for the blue band.

The behavior is rather different for heavy mass irradiations (Fig. 8.2b). For this later case the two bands show an initial rapid growth with fluence up to a maximum (around 10^{12}–10^{13} cm^{-2}) followed by a slower decrease in yield that approaches a

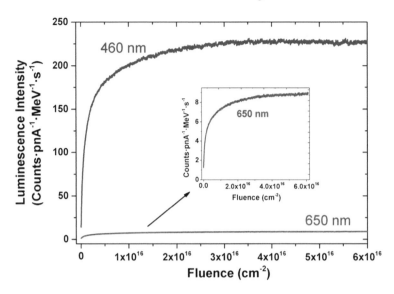

Fig. 8.3 Evolution with fluence of the heights of the 2.7 and the 1.9 eV IL emission bands for RT irradiations with H^+ ions at 2 MeV at fluences up to 6×10^{16} cm^{-2}. The scale for the yield of the red band has been strongly reduced in the inset to facilitate observation

steady level. The fluence corresponding to that maximum yield is comparable for the blue and red emissions. Note for comparison that for H there is no indication of a maximum, neither in the red nor in the blue emission yields, up to fluences of the order of 6×10^{16} cm^{-2} for the IR samples. In the case of irradiations with heavier ions, starting from C, the maximum is always clearly observed in our low-OH samples. It is interesting to remark that the decreasing evolution of the IL yield after the maximum occurs in coincidence with the saturation of the NBOH and ODC center concentrations measured by optical absorption [2]. This indicates that the decrease of the emission yields is not related to a reduction in the concentration of the color centers, acting as recombination centers (see Sect. 8.3).

It is worthwhile for our discussion to evoke the plot of the fluence at which the maximum of the IL yield appears (Φ_{Max}) as a function of the electronic stopping power for the IR silica, Fig. 8.4. This figure clearly illustrates how the position of that maximum shifts to higher fluences as the ion mass is reduced. Note that the scale for the ordinate axis (fluence at which the maximum yield is observed) is logarithmic due to the large span for the magnitude of the effect. In fact, for H and He irradiations the maximum does not appear in our investigated fluence range. These data suggest that the IL kinetics may be modified as a consequence of the structural damage induced by the heavy ion irradiation. As it has been advanced in Sect. 7.3, the structural (macroscopic) effect of SHI irradiation has been characterized by Awazu et al. [3] by using infrared (IR) absorption spectroscopy to monitor the structural changes of silica through the frequency of the first-order ω_4 vibrational mode at the frequency $\omega_4 = 1078$ cm^{-1}. This corresponds to the asymmetric stretching of

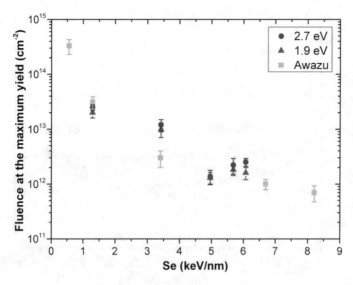

Fig. 8.4 Fluence at which a maximum emission yield is observed (Φ_{Max}) as a function of the electronic stopping power of the projectile ion for silica samples containing $\sim$13 ppm of OH groups. The data obtained in [3] on the irradiation-induced changes in the IR frequency of the ω_4 vibrational mode are included in the plot

the bond, related to the $Si - O - Si$ bond angle, and ultimately responsible for the size of the SiO rings. Similar results have been more recently obtained by Saavedra et al. [4, 5] using reflectance spectroscopy. It has been observed that ω_4 remains constant at low fluences, but experiences a rapid decrease above a certain critical fluence, which markedly decreases on lowering the electronic stopping power of the projectile ion (see Fig. 7.9 on p. 110). This frequency shift accounts for the decrease in the $Si - O - Si$ bond angle and, therefore, the compaction of the silica network. It is, then, useful to quantitatively compare our IL kinetic data with those reported by Awazu et al. [3] (also included in Fig. 8.4) on the fluence at which the frequency ω_4 of the IR mode starts to decrease as a consequence of the structural disorder. The two set of data (IR and IL) confirm that the fluence yielding maximum IL emission rates in Fig. 8.2 correlate quite well with the fluence at which the IR bond frequencies start to be significantly modified.

Another relevant piece of information is the evolution with fluence of the ratio between the yields of the red and blue emissions, $\rho = Y_{RED}/Y_{BLUE}$, that is illustrated in Fig. 8.5 for several irradiation cases. As it will be shown in Sect. 8.2, this ratio cancels the role of the parameters describing the migration of the carriers of the excitation (STEs) and so, it directly conveys information on the concentration of the operative recombination centers (NBOH and ODC). The data in Fig. 8.5 confirm that the yield ratio ρ is lower than 1 and, within the dispersion of experimental data,

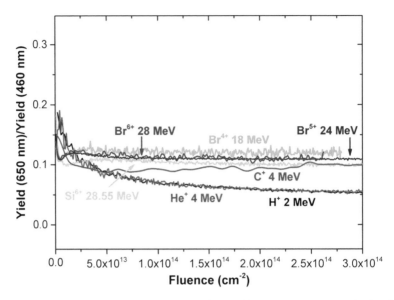

Fig. 8.5 Ratio of the intensities of the two emission bands ($\rho = Y_{RED}/Y_{BLUE}$) for irradiation with the ions considered in this work

is roughly independent of the irradiation fluence and the ion mass. However, this behavior cannot be extrapolated to fluences higher than 10^{16} cm^{-2} (see Fig. 8.3 for irradiation with protons) very likely due to the dominant contribution of the color centers produced via elastic collisions.

8.2   Physical Modeling of STE Dynamics and IL Mechanisms

In order to discuss the effect of electronic stopping power on the kinetics of the IL emission bands a rather detailed analysis of the processes occurring just after the ion bombardment is necessary. The three key elements to describe the IL output are: the nature of the electronic excited states, their migration behavior through the silica network and the effective recombination centers for light emission. Our proposal is to consider that self-trapped excitons (STEs) form readily from the excited electron-hole (e-h) pairs generated by SHI irradiation on SiO$_2$ and are the agents responsible for the transport of the excitation energy. This type of model has been used by Costantini et al. [6] to understand some features of the STE intrinsic blue luminescence at around 2.7 eV induced by ion-beam irradiation at LNT on natural quartz. In our work we are, on the other hand, concerned with the kinetics of the extrinsic luminescence bands induced at RT by SHI irradiation. The existence of excitons, their self-trapping and the electronic structure of STEs in SiO$_2$ have been extensively investigated [7–10] and found responsible for an intrinsic optical transition at around 2.2 in silica (around 2.7 eV in quartz), roughly coinciding with the emission band measured at LNT after synchrotron [11], electron [12] and ion-beam irradiation [6]. A correlated optical absorption at 5.2 and 4.2 eV has also been measured [7] for the STE. The average number (N_{e-h}) of electron-hole pairs generated by every single ion impact along the whole trajectory can be approximately written as:

$$N_{e-h} = \frac{E}{I} \cong \frac{E}{2.5E_g} \tag{8.1}$$

where E is the ion energy, I the effective ionization energy and E_g the band-gap energy of the dielectric material. In Eq. (8.1) we have used an approximation for I, that has been primarily developed to deal with scintillator detectors [13, 14]. A main problem to obtain the number of STEs formed from the N_{e-h} pairs is that reliable information on the self-trapping efficiency for the free e-h pairs is not available. For ionizing irradiation (UV light, X-rays and electrons) it has been estimated [7, 11, 12] to be between 0.1 and 0.01. For ion-beam irradiations, it is well ascertained that only a certain fraction of the generated STEs around the projectile ion trajectory survive the high excitation density and temperature reached at the thermal spike [15]. Therefore, one should multiply expression (8.1) by a surviving factor β that determines the fraction of STEs remaining after the spike and are suitable for recombination and light emission. This factor is known to depend on the type of ion (mass) and its energy E, i.e., the electronic stopping power S_e and, apparently, results from non-radiative e-h recombination processes occurring at the core of the damage track produced by every single projectile ion [15]. It reaches a maximum value for protons (possibly similar to that estimated for ionizing radiation), but it markedly decreases as the ion mass increases. Therefore, as an example, for H at 2 MeV the number of generated and surviving STEs in silica per incident ion, $\beta E/2.5E_g$, is roughly around 10^3, over the whole length of the ion trajectory.

It is expected that the competition between the generation and recombination processes of STEs will lead to a steady population of STEs (N_{STE}) under continuous irradiation. At the start of irradiation of a pristine (pure) sample with a low concentration of recombination centers, one may assume that this steady concentration is determined by the total STE decay time τ. This decay time is 10^{-3} s at or below LNT (radiative lifetime), but decreases down to below 10^{-5} s at RT, due to the nonradiative contributions [7, 16]. It is expected that the lifetime may decrease smoothly on irradiation due to the increase in the number of recombination centers, but we can accept the initial lifetime value as a reasonable assumption.

Then, N_{STE} can be roughly estimated through the expression:

$$N_{STE} = \beta N_{e-h}\phi\tau = \beta \frac{E}{2.5E_g}\phi\tau \tag{8.2}$$

ϕ being the ion flux and τ the total (radiative plus non-radiative) lifetime of the STE. One should note that such steady concentration is reached within an irradiation time of τ (i.e., a fluence of $\phi \cdot \tau$). For typical fluxes $\phi \sim 10^{11}$ cm^{-2}·s^{-1}, this is achieved very early after the start of irradiation, before a significant number of irradiation-induced recombination centers is generated.

In accordance with the available evidence [6, 7], the STEs generated by irradiation move by thermally-activated hoping through the SiO$_2$ network during their lifetime ($<10^{-5}$ s at RT). Those surviving to their intrinsic recombination inside tracks should, finally, recombine, either at pre-existing defects or color centers generated by the irradiation, such as NBOH centers (responsible for the red emission) or ODCII centers (emitting blue light). The intrinsic recombination channel of the STEs does not make a significant contribution to our experimental data at RT since all emission yields (Fig. 8.2) arise from zero at the start of irradiation (pristine sample). In fact, the intrinsic emission yield, per unit time and unit cross-section, is:

$$Y_{int} = \frac{\beta E \phi}{2.5 E_g} \frac{\tau}{\tau_R} \tag{8.3}$$

$\tau_R = 10^{-3}$ s being the radiative lifetime of the STE optical transition. Although reliable values for β are not available, it is likely that for the fluxes used in our experiments that yield is close or below the detection limit of our set-up. The above arguments justify the apparent contradiction between the two alternative models for the blue emission in IL experiments. At LNT and short (pulsed) irradiation, not causing sufficient color centers, one can see [6] the intrinsic emission of the STE, whereas in experiments at RT one should be mostly dealing with extrinsic emissions [1, 17]. Anyhow, the intrinsic recombination channel, even not observable, is, indeed, quite relevant to determine the concentration of STEs available for extrinsic recombination at color centers. Therefore, our proposal is that during irradiation a competition is established between those surviving STEs that recombine at NBOHs, yielding the red emission, or at ODCII centers, emitting the blue light. This competition between the red and blue emission channels should be responsible for the observed shapes of the kinetic curves for the two IL bands as a function of fluence.

In accordance with the above view the yields of the red (R) and blue (B) emissions per unit time and unit cross-section can be, respectively written as:

$$Y_{R,B}(\Phi) = \alpha_{R,B} R_{R,B}(\Phi) N_{STE} \eta_{R,B} \tag{8.4}$$

Being Φ the irradiation fluence. α_R and α_B are detection efficiency factors for the emitted photons that depend on experimental geometry and detectors sensitivity. $R_{R,B}$ represents the number of red (NBOH) and blue (ODCII) recombination centers that each STE meets per unit time during its migration through the silica network within their lifetime. $\eta_{R,B}$ stands for a factor including the trapping cross-sections and quantum efficiencies for the optical transitions responsible for the light emission.

In order to estimate the evolution with fluence for the extrinsic emissions yields Y one should dig into the dependence of the R_R and R_B factors with fluence. A rigorous analysis of the processes is out of the scope of this thesis, and would require more sophisticated tools such as MonteCarlo simulations of the STE migration and trapping at recombination sites [18]. However, as a simple approximation, one may assume a low enough concentration for the two types of recombination centers and ignore repeated visits of the STE to the same site. Then, a rough expression for the R factors is:

$$R_{R,B}(\Phi) = \nu c_{R,B}(\Phi) = \nu_0 e^{-\varepsilon/kT} c_{R,B} \tag{8.5}$$

ν being the jump frequency for the STE between adjacent lattice sites and ν_0 a suitable pre-exponential factor around 10^{12} s^{-1}. $c_{R,B}$ stands for the (fluence dependent) relative atomic concentrations of NBOH and ODCII centers and ε is the energy barrier for hopping. This energy ε for the hopping has been estimated to be around 0.15–0.18 eV [6], which essentially corresponds to that for self-trapped holes. It leads to a hopping rate ν of 7.5×10^8 s^{-1} at RT, i.e., to less than 10^4 jumps or lattice sites visited during the STE lifetime. In other words, the migration length or average distance traveled by the STE during its lifetime is $L_{STE} = \nu \cdot a \cdot \tau$, a being the jump length. In the initial linear stage of color center growth, the relative atomic concentrations are in all cases $c < 10^{-4}$ and so the rate of STE trapping at color centers during the lifetime can be considered small enough to guarantee the constancy of lifetime and justify our low concentration approximation in Eq. (8.5). Therefore, one may expect that τ is independent of fluence (i.e., of the irradiation induced color center concentrations) so that N_{STE} can be taken constant, at least during the initial stage of irradiation (before reaching saturation). This accounts for the roughly linear initial growth of the two luminescence yields with fluence, starting from zero.

8.3 Physical Discussion of the Experimental Results: Role of Network Straining

In base to the above theoretical analysis one can, now, discuss the shape of the kinetic curves for the two bands under ion-beam irradiation with either light or heavy ions, and the occurrence of a maximum in the IL kinetics for the case of heavy ions. In principle, one would expect from Eq. (8.1) that the IL yield curves would correlate with those for the color center growth. This is the behavior found for our H irradiations as previously noted by Nagata et al. [1, 17]. The behavior of SHI irradiations is rather different, i.e., the IL yields show a pronounced decrease after an initial fast growth up to a maximum level. Therefore, they diverge dramatically from the growth curves for the absorption of the relevant color centers [19, 20]. In fact, the decreasing IL rates found after the maximum yield, once the color center (recombination center) concentrations have reached steady saturation, are not easy to explain. Therefore, one has to think of some processes affecting the R factors which determine the yields in Eq. (8.4). It is known that the electronic excitations associated to SHI irradiations cause not only color center formation but, also, strong distortions of the SiO_2 network, leading to macroscopic compaction. A number of proposals have been advanced for those phenomena involving $Si - O - Si$ bond scission by the electronic excitations as a primary mechanism for color center formation [7, 8, 11, 21, 22] (microscopic damage). This scission, followed by bond healing and reordering, induces significant changes in the network structure and in the ring size distribution that have been studied by IR [3] and Raman [4] spectroscopy. As a net effect, smaller ring sizes are favored (compaction), that induce changes in the macroscopic properties such as refractive index [23, 24] and optical absorption [25] near the edge (macroscopic damage). As it was shown in Sect. 7.4.1 there are two main scission mechanisms leading to the formation of NBOH and ODCII centers (Sect. 7.4.1, p. 109), that act as recombination centers for STEs and so to light emission [16, 22, 25, 26]. In principle, the evolution of the concentration of such centers could be monitored by in-situ ionoluminescence experiments and this will be the objective of future work.

Anyhow, the experimental data in Fig. 8.4 show that the structural changes measured by IR spectroscopy [3] can be well correlated with the kinetic behavior of the IL emissions. In particular, the maximum of the IL, implying a dramatic reduction in emission yields, occurs at fluences at which a strong decrease in the vibrational frequency ω_4 starts to be observed. Within our IL model the reason may be associated to changes in the STE migration dynamics, either the STE lifetime and/or their mobility, i.e., the product $\nu \cdot \tau$ or the migration length L_{STE}. If L_{STE} is reduced as a consequence of compaction one can understand that the two emission yields experience a correlated decrease due to the reduced probability for the STE to meet a recombination center during its lifetime. In other words, the quantitative comparison offered in Fig. 8.4 appears to provide a sound support for the physical basis of our model. Obviously, additional experiments on STE migration behavior and theoretical efforts have to be carried out to understand the details of the model.

A robust parameter to describe the evolution of the IL is that referring to the yield ratio $\rho = Y_{RED}/Y_{BLUE}$. From Eqs. (8.1) and (8.5), one obtains:

$$\rho = \frac{Y_R}{Y_B} = \frac{\alpha_R c_R \eta_R}{\alpha_B c_B \eta_B} \tag{8.6}$$

where some parameters, as the STE population and the STE transport parameters, have been canceled. Therefore, the only factors depending on fluence are the concentration of the corresponding recombination centers. Considering that α_R/α_B, and η_R/η_B are constant and independent of the irradiation fluence, it comes out that ρ (i.e., $\rho(\Phi)$) is only determined by the evolution of the concentrations of the NBOH and ODC centers: $\rho \propto c_R/c_B$. Although the data in Fig. 8.5 are pending of a more detailed analysis, the approximate constancy of the yield ratio and its independence electronic stopping power for SHI ions are consistent with electronic scission mechanisms represented by the competing reactions Eqs. 7.2 and 7.3 on p. 110. For the case of light projectile ions (H, He) this rule is broken and the production of ODC defects becomes strongly enhanced. In summary, the IL experiments are expected to give in real time the ratio of the concentrations for the two types of color centers,

NBOH and ODCII, acting as recombination centers. Therefore, the IL data should be quite relevant in the discussion of the defect formation mechanisms, which should be the objective of future work.

8.4 Mathematical Formulation of the IL Emission Kinetics: Damage Cross-Sections

Although it is, possibly, premature one may propose a simple mathematical formulation for the red and blue emission kinetics. Let us use a simple random-walk approach and assume that the atomic fraction for the NBOHC and ODC concentration is $c_{R,B} \ll 1$, so that the repeated visiting of a given lattice site can be neglected. Moreover, one may consider a Poisson kinetics for the growth of color center concentration:

$$c_{R,B}(\Phi) = c_{R,B}(\infty) \left(1 - e^{-\sigma_{R,B}\Phi}\right) \qquad (8.7)$$

$\sigma_{R,B}$ being the cross-section for the radiation-induced creation of NBOH and ODC centers respectively (microscopic damage cross-section) and $c_{R,B}(\infty)$ the color center concentration at saturation. Then, Eq. 8.4 writes:

$$Y_{R,B}(\Phi) = \alpha_{R,B} \cdot N_{STE} \cdot \eta_{R,B} \cdot c_{R,B}(\infty) \cdot v(\Phi) \cdot \left(1 - e^{-\sigma_{R,B}\Phi}\right) \propto L_{STE}(\Phi) \cdot \left(1 - e^{-\sigma_{R,B}\Phi}\right) \qquad (8.8)$$

The dependence on fluence of the various parameters has been explicitly indicated. Therefore, $Y_{R,B}$ is the product of a factor, $1 - e^{-\sigma_{R,B}\Phi}$, that follows the rapid growth of the NBOH or ODC concentrations during the first stage of the kinetics curves and a factor proportional to $L_{STE}(\Phi)$ $(L_{STE}(\Phi) = v \cdot \tau \cdot a)$. The fluence dependence of this second factor is proposed to be responsible for the dramatic reduction in yield during the second stage. This dependence may be associated to the jump frequency, through the variation in the barrier energy ε with the lattice disorder caused by irradiation, or alternatively to changes in the STE lifetime. Presently, the available information does not permit to decide on those two possibilities. Consequently, the evolution of L with fluence cannot be theoretically quantified, although, according to the main proposal of our model, it is expected to be a decreasing function of Φ. It should vary from an initial value L_0 (at low fluences) to a final value L_∞ (at high fluences and high lattice disorder). For simplicity one can assume a simple exponential decay:

$$L_{STE}(\Phi) = L_\infty + (L_0 - L_\infty) e^{-\Sigma_A \Phi} \qquad (8.9)$$

Σ_A being an amorphization cross-section that characterizes the rate at which macroscopic disorder (macroscopic damage) is introduced by the irradiation. It is expected to be much lower than the cross-sections $\sigma_{R,B}$ for the growth of NBOH and ODC concentrations. For simplicity one can consider a limit realistic case, $L_\infty \approx 0$,

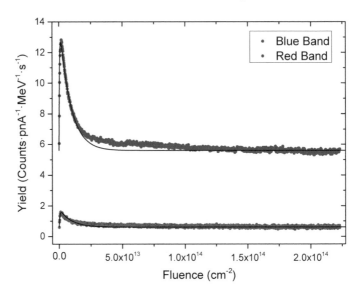

Fig. 8.6 Example of the fits (black solid lines) of the red and blue IL emission kinetics in the case of 28 MeV Br^{6+} irradiation in silica. These curves were obtained by fitting Eq. (8.10) to the IL experimental data

i.e., the STE migration is fully inhibited at high fluences. Then, it comes out that the kinetics for the red and blue emissions consist of a saturating exponential describing the growth of the NBOH/ODC center concentration multiplied by a decreasing exponential, describing the structural damage caused by the irradiation:

$$Y_{R,B}(\Phi) = k \cdot \left(1 - e^{-\sigma_{R,B}\Phi}\right) e^{-\Sigma_A \Phi} + k' \tag{8.10}$$

The parameters k and k' depend on the experimental setup. The two exponential factors in Eq. (8.10) should account for the two stages of the red and blue emission kinetics. Some preliminary fits of the IL kinetics using this equation have been done for the case of 28 MeV Br^{6+} ions in silica. The curves obtained with the fits are shown in Fig. 8.6 for the red and blue bands. The values obtained for the microscopic and macroscopic cross-sections are: $\sigma_R \approx 2.0 \times 10^{-12}$ cm^2, $\sigma_B \approx 1.9 \times 10^{-12}$ cm^2, and $\Sigma_A \approx 1.2 \times 10^{-13}$ cm^2. This is only an example to show the validity of the model and of the mathematical expressions obtained here, more detailed fits with other ions and energies could be done in the future using Eq. (8.10) to determine the values of the cross sections and their dependence on the stopping power of the incident ions.

References

1. S. Nagata, S. Yamamoto, K. Toh, B. Tsuchiya, N. Ohtsu, T. Shikama, H. Naramoto, Luminescence in SiO_2 induced by MeV energy proton irradiation. J. Nucl. Mater. **329–333**(B), 1507–1510 (2004). Proceedings of the 11th International Conference on Fusion Reactor Materials (ICFRM-11)

2. O. Peña-Rodríguez, D. Jiménez-Rey, J. Manzano-Santamaría, J. Olivares, A. Muñoz, A. Rivera, F. Agulló-López, Ionoluminescence as sensor of structural disorder in crystalline sio2: determination of amorphization threshold by swift heavy ions. Appl. Phys. Express **5**(1), 011101 (2012)

3. K. Awazu, S. Ishii, K. Shima, S. Roorda, J.L. Brebner, Structure of latent tracks created by swift heavy-ion bombardment of amorphous sio2. Phys. Rev. B: Condens. Matter Mater. Phys. **62**, 3689–3698 (2000)

4. R. Saavedra, M. León, P. Martín, D. Jiménez-Rey, R. Vila, S. Girard, A. Boukenter, Y. Querdane, Raman measurements in silica glasses irradiated with energetic ions. AIP Conf. Proc. **1624**, 118–124 (2014)

5. R. Saavedra, P. Martín, D. Jimenez-Rey, R. Vila, Structural changes induced in silica by ion irradiation observed by IR reflectance spectroscopy. Fusion Eng. Design **98–99**, 2034–2037 (2015)

6. J.M. Costantini, F. Brisard, G. Biotteau, E. Balanzat, B. Gervais, Self-trapped exciton luminescence induced in alpha quartz by swift heavy ion irradiations. J. Appl. Phys. **88**, 1339–1345 (2000)

7. A.K.S. Song, R.T. Williams, *Self-trapped Excitons* (Springer, Berlin, 1996)

8. N. Itoh, A.M. Stoneham, *Materials Modification by Electronic Excitation* (Cambridge University Press, Cambridge, 2001)

9. S. Ismail-Beigi, S.G. Louie, Self-trapped excitons in silicon dioxide: mechanism and properties. Phys. Rev. Lett. **95**, 156401 (2005)

10. R.M. Van Ginhoven, H. Jónsson, L.R. Corrales, Characterization of exciton self-trapping in amorphous silica. J. Non-Crystall. Solids **352**(23–25), 2589–2595 (2006)

11. F. Messina, L. Vaccaro, M. Cannas, Generation and excitation of point defects in silica by synchrotron radiation above the absorption edge. Phys. Rev. B: Condens. Matter Mater. Phys. **81**, 035212 (2010)

12. C. Itoh, K. Tanimura, N. Itoh, Optical studies of self-trapped excitons in SiO_2. J. Phys. C: Solid State Phys. **21**(26), 4693–4702 (1988)

13. F.S. Goulding, Y. Stone, Semiconductor radiation detectors: basic principles and some uses of a recent tool that has revolutionized nuclear physics are described. Science **170**(3955), 280–289 (1970)

14. R.C. Hughes, Charge-carrier transport phenomena in amorphous SiO_2: direct measurement of the drift mobility and lifetime. Phys. Rev. Lett. **30**, 1333–1336 (1973)

15. K. Michaelian, A. Menchaca-Rocha, Model of ion-induced luminescence based on energy deposition by secondary electrons. Phys. Rev. B: Condens. Matter Mater. Phys. **49**, 15550–15562 (1994)

16. B.J. Luff, P.D. Townsend, Cathodoluminescence of synthetic quartz. J. Phys.: Condens. Matter. **2**, 8089–8097 (1990)

17. S. Nagata, S. Yamamoto, A. Inouye, B. Tsuchiya, K. Toh, T. Shikama, Luminescence characteristics and defect formation in silica glasses under h and he ion irradiation. J. Nucl. Materi. **367–370**(B), 1009–1013 (2007). Proceedings of the Twelfth International Conference on Fusion Reactor Materials (ICFRM-12)

18. B. Sturman, M. Carrascosa, F. Agulló-López, Light-induced charge transport in $LiNbO_3$ crystals. Phys. Rev. B: Condens. Matter Mater. Phys. **78**, 245114 (2008)

19. M. Ma, X. Chen, K. Yang, X. Yang, Y. Sun, Y. Jin, Z. Zhu, Color center formation in silica glass induced by high energy fe and xe ions. Nucl. Instrum. Methods Phys. Res. Sect. B: Beam Interact. Mater. Atoms **268**(1), 67–72 (2010)

20. J. Manzano-Santamaría, J. Olivares, A. Rivera, O. Peña-Rodríguez, F. Agulló-López, Kinetics of color center formation in silica irradiated with swift heavy ions: thresholding and formation efficiency. Appl. Phys. Lett. **101**, 154103 (2012)
21. H. Hosono, H. Kawazoe, N. Matsunami, Experimental evidence for Frenkel defect formation in amorphous SiO_2 by electronic excitation. Phys. Rev. Lett. **80**, 317–320 (1998)
22. K. Kajihara, M. Hirano, L. Skuja, H. Hosono, Intrinsic defect formation in amorphous SiO_2 by electronic excitation: bond dissociation versus Frenkel mechanisms. Phys. Rev. B: Condens. Matter Mater. Phys. **78**, 094201 (2008)
23. J. Manzano, J. Olivares, F. Agull'o-L'opez, M.L. Crespillo, A. Moroño, E. Hodgson, Optical waveguides obtained by swift-ion irradiation on silica (a-sio2). Nucl. Instrum. Methods Phys. Res. Sect. B: Beam Interact. Mater. Atoms **268**(19), 3147–3150 (2010). Radiation Effects in Insulators, Proceedings of the 15th International Conference on Radiation Effects in Insulators (REI), 15th International Conference on Radiation Effects in Insulators (REI)
24. O. Peña-Rodríguez, J. Manzano-Santamaría, J. Olivares, A. Rivera, F. Agulló-López, Refractive index changes in amorphous SiO_2 (silica) by swift ion irradiation. Nucl. Instrum. Methods Phys. Res. Sect. B: Beam Interact. Mater. Atoms **277**, 126–130 (2012)
25. H. Hosono, Y. Ikuta, T. Kinoshita, K. Kajihara, M. Hirano, Physical disorder and optical properties in the vacuum ultraviolet region of amorphous SiO_2. Phys. Rev. Lett. **87**, 175501 (2001)
26. T.E. Tsai, D.L. Griscom, Experimental evidence for excitonic mechanism of defect generation in high-purity silica. Phys. Rev. Lett. **67**, 2517–2520 (1991)

Part III
Ion-Irradiation Damage in MgO

Chapter 9
MgO Under Ion Irradiation at High Temperatures

The damage process in magnesia (MgO) single crystals irradiated with 1.2 MeV Au^+ ions at high temperatures and at different fluences (up to 4×10^{14} cm^{-2}) has been studied. For this study two different techniques have been used: Rutherford Backscattering Spectrometry in Channeling configuration (RBS/C) and High-Resolution X-Ray Diffraction (HRXRD).

The samples were irradiated at 573, 773 and 1073 K with Au^+ ions using the ARAMIS tandem accelerator of the CSNSM in Orsay. During irradiations, the samples were tilted by an angle of 7° relatively to the <100> main axis; the edges of the samples were not cut along a particular direction. In these conditions, as demonstrated in [1], channeling of Au ions is minimized. A beam raster system was used to ensure uniform ion irradiation, and, in order to prevent any sample overheating, the ion flux did not exceed 2.5×10^{11} cm$^{-2} \cdot$s^{-1}.

Au^+ ions at 1.2 MeV were used in order to have a preponderant contribution of ballistic effects on the damage creation process. The fluences used were in the range 5×10^{12}–4×10^{14} cm^{-2}. Based on SRIM calculations [2, 3], the mean projected range of Au particles has been estimated to be $R_p \sim 183$ nm with a range straggling of $\Delta R_p \sim 33$ nm (Fig. 9.1a). The maximum electronic energy loss, from both ions and recoils, was $S_e \sim 4$ keV/nm, and the maximum nuclear energy loss was $S_n \sim 5$ keV/nm (Fig. 9.1b). The conversion factor for the displacements per atom (*dpa*) at the damage peak, using a threshold displacement energy of $E_d = 55$ eV for both Mg and O sublattices (see [4] and references therein), is ~ 2.7x10^{15} dpa·cm^{-2}; the corresponding *dpa* values are hence in the range between ~ 0.01 and ~ 0.5.

Section 9.1 presents a general overview of the whole damage accumulation process in MgO in the fluence range from 5×10^{12} up to 4×10^{14} cm^{-2}. Using the RBS technique, two steps have been found to occur during this process. XRD reciprocal space maps confirm the existence of these two steps and provide some information about the main features of each step. Then, in Sect. 9.2, the initial stages of the damage process in MgO are studied more finely using the XRD technique, which offers information about the defect concentration.

© Springer Nature Switzerland AG 2018
D. Bachiller Perea, *Ion-Irradiation-Induced Damage in Nuclear Materials*,
Springer Theses, https://doi.org/10.1007/978-3-030-00407-1_9

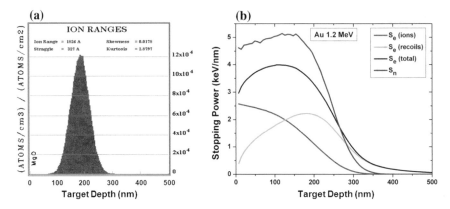

Fig. 9.1 SRIM calculations for 1.2 MeV Au$^+$ ions in MgO. **a** Ion distribution. **b** Nuclear (S_n) and electronic (S_e) stopping powers

9.1 Full Damage Accumulation Process in MgO Irradiated with MeV Au Ions at Elevated Temperatures

In this section, we make use of the RBS/C technique to obtain the disorder depth profiles and damage accumulation in single-crystalline MgO samples irradiated at 573, 773, and 1073 K over a large ion fluence range (more precisely, until saturation of the disorder). Additional X-ray diffraction data are provided to support the RBS/C results.

Irradiated crystals were analyzed by RBS/C using the tandem accelerator of the CSNSM. A 1.4 MeV ^{4}He$^+$ ion beam was used, and a Si detector was placed at a 165°}scattering angle. The detector resolution was on the order of 15 keV, which corresponds to a depth resolution of $\sim$10 nm at the sample surface. Prior to RBS/C analyses, a thin carbon layer was deposited on the sample surface to avoid charging effects. Simulation of the RBS/C spectra was achieved using the McChasy Monte-Carlo code [5], with the assumption that the irradiation-induced disorder can be represented as a fraction of randomly displaced atoms, f_D.

XRD reciprocal space maps were recorded on irradiated crystals using the Philips X'Pert PRO MRD diffractometer at the CTU-IEF in Orsay; both the equipment and the procedure are described in [6] and in Sect. 5.3. The formalism is presented in [7, 8]; briefly, (i) K_N and $K_\parallel$ are the normal (out-of-plane) and parallel (in-plane) components of the scattering vector $\vec{K}$ ($|\vec{K}| = 2sin\theta/\lambda$), respectively; (ii) $\vec{H}_{(004)}$ refers to the reciprocal lattice vector for the (004) reflection; (iii) $\vec{q}_N$ is defined as $\vec{K}_N - \vec{H}_{(004)}$ and represents the deviation from the reciprocal lattice vector. In the following, positions on the maps are located by K_N and $K_\parallel$, but also by, respectively, (i) $\left(-q_N/H_{(004)}\right)$, which is equal to the elastic strain in the direction normal to the surface of implanted samples, and (ii) $\left(\Delta K_\parallel/H_{(004)}\right)$, which directly gives the width (here in degrees) of the reciprocal lattice point in the transverse direction.

9.1.1 Disorder Depth Profiles

Figure 9.2 displays spectra recorded in random (stars) and aligned geometries for MgO single crystals irradiated at several ion fluences and at the three tested temperatures: 573 K (a), 773 K (b) and 1073 K (c). The spectrum in the random geometry presents two plateaus (below 700 and 500 keV) that are due to the backscattering of the analyzing particles (He ions) from Mg and O atoms, respectively. The spectrum recorded along the <100> axis on a pristine crystal presents a very low value of the yield ($\chi_{min} \sim 0.025$), which attests to the good quality of the MgO single crystals. Irrespective of the temperature, the spectra recorded in channeling configuration on irradiated crystals exhibit an increase of the backscattering yield with increasing ion fluence.

In order to get quantitative information on the irradiation-induced disorder, channeling data were analyzed with the McChasy Monte Carlo simulation code [5]. Fits (represented as solid lines in Fig. 9.2) of the RBS/C data provide the depth distributions of the disorder (f_D) for both Mg and O sublattices. Figure 9.3 displays the damage profiles in the Mg cationic sublattice for the three temperatures; the O anionic sublattice follows a similar trend. As suggested by the examination of the raw data, the disorder increases with ion fluence whatever the irradiation temperature is. For all temperatures, the damaged thickness extends from the surface to ~600 nm (note that part of the tail in the disorder profile at large depth is due to the fact that for the simulations we used the crude assumption that only randomly displaced atoms are present).

A shift of the damage peak toward greater depth is visible at increasing fluence (Fig. 9.3), as already observed in recent works for RT irradiations with 1 MeV [1] and 1.2 MeV Au$^+$ ions [9]. This shift is however not the same for the three temperatures. To support this finding, in Fig. 9.4, the depth of the damage peak is plotted as a function of the Au fluence for the three irradiation temperatures. It clearly appears that it varies with both parameters. At low fluence, this peak is located almost at the same depth, ~140 nm, at 573 and 773 K, whereas it is at ~220 nm at 1073 K. With increasing fluence, the damage peak shifts and reaches the same depth location, ~280 nm, at 573 and 773 K. Note that this depth is identical to that determined at the same fluence (4×10^{14} cm^{-2}) for RT irradiation [9]. The damage peak is found to be deeper at 1073 K (~310 nm).

9.1.2 Damage Accumulation

The damage accumulation was determined by plotting the maximum of the damage fraction (f_D^{max}) as a function of the Au ion fluence for the three investigated temperatures (Fig. 9.5). Note that these f_D^{max} values were extracted from Fig. 9.3 by always taking the maximum damage fraction irrespective of its depth location. It readily appears that these disordering curves cannot be fitted with a single impact approach.

Fig. 9.2 RBS/C spectra of
MgO crystals irradiated with
1.2 MeV Au$^+$ ions at
different temperatures and
fluences. Solid lines
correspond to the fits
obtained with the McChasy
code

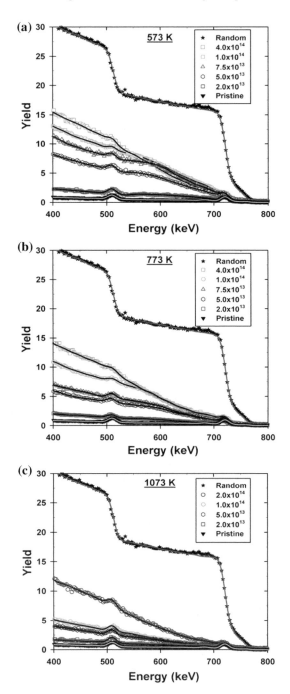

Fig. 9.3 Disorder depth
profiles obtained, for the
three studied temperatures,
by fitting the RBS/C spectra
of Fig. 9.2 with the McChasy
simulation code [5]

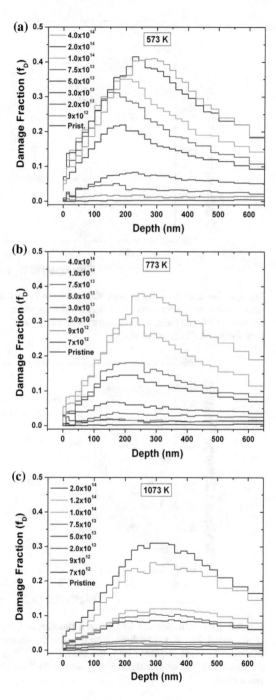

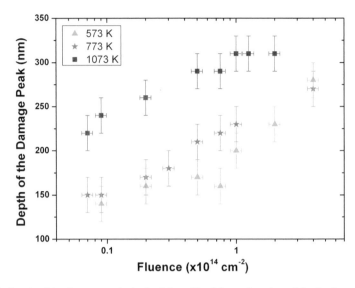

Fig. 9.4 Depth of the damage peak obtained from Fig. 9.3 as a function of the Au fluence for the three irradiation temperatures. Note that the fluence scale is logarithmic. The relative error in the fluence is 10%, and the error in the depth is 20 nm, which corresponds to the thickness of the layers used for the McChasy simulations

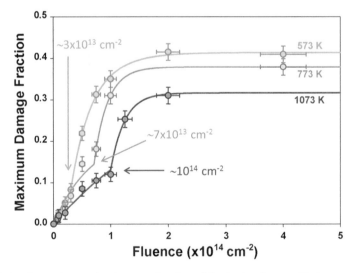

Fig. 9.5 Maximum damage fraction as a function of the Au ion fluence. Circles correspond to experimental data and solid lines represent the fits obtained using the MSDA model [10]. The relative error in the fluence is 10%, and the error in f_D^{max} has been calculated taking into account the incertitude on the cumulated charge of the RBS spectra and the precision of the McChasy simulations

We thus used the Multi-Step Damage Accumulation (MSDA) model [10] to fit these data. This phenomenological model relies on the assumption that the disorder accumulation occurs in several, distinct n steps. In the present case, the curves were fitted using $n = 2$, which means that two steps were identified in the disordering process. This result is in perfect agreement with a previous work on RT irradiation experiments in MgO [9]. Another piece of information provided by the fitting is that the damage level saturates at high fluence for the three temperatures. Nevertheless, an effect of the temperature is evidenced, as this saturation level slightly, but clearly, decreases with increasing temperature, from ∼0.4 at 573 K to ∼0.3 at 1073 K. This low value of the saturation level indicates that the MgO crystalline structure does not experience an amorphous transformation under heavy-ion irradiation at high temperatures (up to a fluence of 4×10^{14} cm^{-2}) and that it is extremely radiation-resistant. The change in irradiation temperature also led to another important modification in the damage accumulation: the higher the temperature, the higher the fluence for the transition between step 1 and step 2. For instance, this transition fluence is ∼3×10^{13} cm^{-2} at 573 K (very close to that at RT [9]), ∼7×10^{13} cm^{-2} at 773 K and ∼10^{14} cm^{-2} at 1073 K (see Fig. 9.5).

The decomposition of the curves in Fig. 9.5 in the two steps is shown in Fig. 9.6 for the three temperatures. It appears that not only the transition fluence is shifted to higher fluences with the temperature, but also the weight of the first step on the whole damage accumulation process increases with temperature.

9.1.3   Discussion

Defect energetics in MgO is rather well known, particularly due to numerous computational works [11–14]. All studies pointed out that Mg and O vacancies are not mobile at RT, with migration energies of a few electronvolts (eV). On the contrary, both interstitials have low migration barriers allowing their migration at RT. For instance, values of 0.32 and 0.40 eV for O and Mg (neutral) single-interstitials were computed [12]. Di-interstitials and tri-interstitials were also found to be mobile at RT [12]. This high defect mobility was put forward to explain the shift of the damage peak toward greater depth observed with increasing fluence upon irradiation at RT [9]. Indeed, during irradiation, defects migrate toward both the surface and the bulk, but they are annihilated at the surface, which leads to an apparent shift in depth of the damage peak. In the actual experiments at elevated temperatures, the depth of the damage peak is also found to increase with the fluence. A different behavior was observed in the case of cubic zirconia (c-ZrO$_2$) [15], where the defect mobility is rather limited as calculated by computer modeling techniques in [16]. Actually, a shift of the damage peak toward the surface (and not toward the bulk) was observed, and only at temperatures higher than 773 K [15]. In MgO, the shift of the peak occurs already at RT, and it is similar at RT, 573 and 773 K. The shift is more important at 1073 K. Vacancies were shown to be mobile at temperatures higher than 873 K [17]; the present result indicate that in this temperature range, the mobility of interstitials,

Fig. 9.6 The red and green dashed lines represent the two steps obtained with the MSDA model

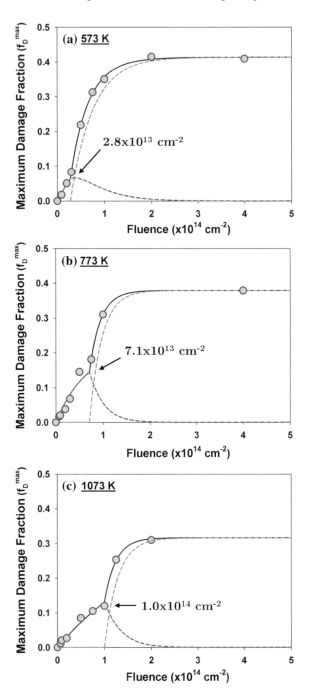

which are the defects mainly detected by RBS/C (and XRD), is also significantly enhanced.

The occurrence of two steps in the damage build-up at RT has been previously ascribed to the formation of two different types of predominant defects [9]: point defects and small defect clusters are generated in the first step, and in the second step their growth leads to the formation of dislocation loops (which are primarily formed at the damage peak where the defect density is the highest). A third step was also observed, but it was due to the fact that the disorder level was followed at a constant depth, and at high fluence, the shift of the peak led to an artificial decrease in the disorder. An analogous disordering process was also reported in a recent work on RT irradiation of MgO in very similar conditions (1 MeV Au^+ ions) [1]. Therefore, the multi-step damage accumulation process is characteristic of MgO for the RT to 1073 K temperature range. This behavior is in contrast with the one observed for other ceramics like, e.g., SiC [18] or $SrTiO_3$ (STO) [19] for which increasing the irradiation temperature leads to a dramatic change in the disordering process. In fact, in these materials, the irradiation-induced amorphisation is prevented and the residual disorder is maintained at a low level due to a very efficient dynamic defect annealing that takes place at moderate temperature ($\sim$623 K for SiC and 400 K for STO). On the contrary, the MgO behavior resembles that of cubic zirconia for which a similar multi-step damage accumulation process was observed for the 80–1073 K range [20]. However, it is surprising to note that in the case of c-ZrO_2, the transition fluence decreases with increasing temperature, in total contrast with the actual finding in MgO.

Increasing the irradiation temperature necessarily leads to an enhanced defect mobility. Then, there must be a competition between defect annealing and defect clustering. In the case of c-ZrO_2, enhanced clustering was assumed, leading to earlier formation of extended defects (and thus earlier occurrence of step 2). In the case of MgO, it seems that defect annealing is favored, delaying the generation of dislocation loops. As for the shift of the damage peak, this delay is found to be the most pronounced at 1073 K. One explanation could lie in the fact that above 873 K, both interstitials and vacancies are mobile [17], favoring defect recombination. Another evidence of defect annealing upon irradiation is given by Fig. 9.7 that presents f_D^{max} as a function of the irradiation temperature for three selected Au fluences: 2×10^{13} cm^{-2}, 7.5×10^{13} cm^{-2} and 2×10^{14} cm^{-2}.

In the three cases shown in Fig. 9.7, a decrease of the disorder level is observed with increasing temperature. For the 2×10^{13} cm^{-2} and 2×10^{14} cm^{-2} fluences, which correspond, respectively, to step 1 and step 2 for the three irradiation temperatures, the decrease is nearly linear. This result is in qualitative agreement with that of Usov et al. which also showed a linear decrease of the disorder with temperature for this temperature range [21]. For the 7.5×10^{13} cm^{-2} fluence, the trend in the disorder recovery is somehow more complex because this fluence corresponds to step 2 at 573 and 773 K, but to step 1 at 1073 K; this difference might explain the change in the decreasing rate at 773 K. However, a clear decrease in the disorder with increasing temperature is also observed for this fluence.

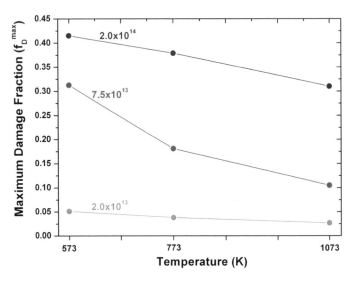

Fig. 9.7 Maximum damage fraction as a function of the irradiation temperature for three different ion fluences

In order to confirm the conclusions drawn from RBS/C data regarding both the multi step damage accumulation process and the temperature-induced defect annealing, we performed XRD reciprocal space maps. These maps are displayed in Fig. 9.8a, b and c correspond to a low irradiation fluence (belonging to step 1) at 573, 773 and 1073 K, respectively and Fig. 9.8d, e and f were recorded for samples irradiated at high fluence (very close to, or at the disorder saturation, i.e., at fluences corresponding to step 2) at the same three temperatures. The comparison of the first set of maps at low fluence indicates that the elastic strain decreases with increasing temperature, as evidenced by the shrinking of the streak along the K_N direction. This finding is related to a decrease in the defect density, which confirms the annealing effect with the rise in temperature. Moreover, the scattered intensity is confined around a narrow $K_\parallel$ region around these streaks that exhibit a fringe pattern. These features are indicative, as already demonstrated in, e.g., c-ZrO$_2$ [22], of the presence of only small defect clusters (except maybe at 1073 K where a spreading of the intensity is visible along $K_\parallel$, which suggests, for some yet unexplained reason, the presence of larger defects). For the second set of maps at high fluence, a featureless streak is observed along the K_N direction, along with a significant spreading of the scattered intensity along the $K_\parallel$ direction. These observations indicate a significant diffuse scattering component that is related to the presence of extended defects [22]. Therefore, at least qualitatively, two steps in the damage build-up are also evidenced by the XRD data. Furthermore, the XRD data strongly suggest that the defects formed in each step at elevated temperatures are similar to those previously observed at RT: in the first step point defects are generated, and they grow in the second step forming dislocation loops.

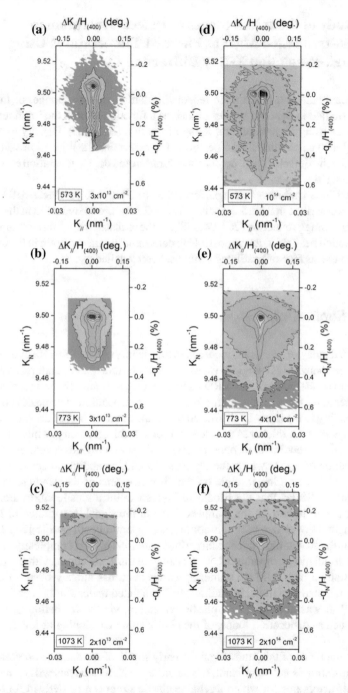

Fig. 9.8 XRD reciprocal space maps recorded on MgO crystals irradiated with 1.2 MeV Au$^+$ ions at different fluences and temperatures: **a** 3x10^{13} cm^{-2} at 573 K, **b** 3x10^{13} cm^{-2} at 773 K, **c** 2x10^{13} cm^{-2} at 1073 K, **d** 10^{14} cm^{-2} at 573 K, **e** 4x10^{14} cm^{-2} at 773 K, and **f** 2x10^{14} cm^{-2} at 1073 K

9.2   Study of the Initial Stages of Defect Generation in Ion-Irradiated MgO at Elevated Temperatures Using High-Resolution X-Ray Diffraction

The lattice strain, and thus the corresponding lattice volume change, of the MgO samples studied in Sect. 9.1 was examined as a function of both the ion fluence and temperature using high-resolution X-ray diffraction (HRXRD). Density Functional Theory (DFT) calculations were performed to determine the relaxation volumes of Mg and O point defects. Then, defect concentration and defect generation efficiencies were evaluated.

The XRD measurements were performed with a Philips X'Pert PRO MRD diffractometer as described in Sect. 5.3. Symmetric $\theta - 2\theta$ scans were recorded in the vicinity of the (400) Bragg reflection ($2\theta \sim 94.058°$) of the rock salt magnesia structure. From the diffraction peak shift, it was possible to determine the elastic strain in the direction normal to the surface of irradiated samples, hereby called ε_N.

9.2.1   Strain Evolution

Figure 9.9 displays $\theta - 2\theta$ scans recorded in the vicinity of the (400) reflection for pristine and irradiated MgO single crystals at the three studied temperatures: 573 K (a), 773 K (b) and 1073 K (c). All XRD curves exhibit the same shape independent of the temperature. First, on the high-angle side, a sharp and intense peak is observed corresponding to the intensity diffracted by the unirradiated part of the samples (note that X-rays probe a thickness of a few microns in this geometry, while the Au ion range is only a few hundred nanometers). This peak provides a reference for the determination of ε_N shown on the top axis. Second, an additional scattering signal is clearly visible at a lower angle on the diffractograms of the irradiated samples. This signal indicates that the irradiated layers exhibit a tensile strain along their surface normal; moreover, the complex shape of the signal, constituted of interference fringes, is evidence of a non-homogeneous strain depth profile [7, 23–25]. It is noteworthy that at high fluence this fringe pattern is almost completely vanishing under all irradiation temperatures. This change in the XRD curve, with the presence of essentially a broad diffuse scattering component, has already been observed in several irradiated materials, including MgO irradiated under similar conditions but at RT [9]. It was attributed to a plastic relaxation, via the formation of extended defects such as dislocation loops, of the elastic strain developing at low fluence [8, 26, 27].

Since the focus of this study is on the early stages of the damage process, where defect clustering is not significant, the quantitative XRD analysis will be restrained to the diffractograms for which a reliable elastic strain can be derived for all three temperatures (i.e., up to a fluence of 7.5×10^{13} cm^{-2}).

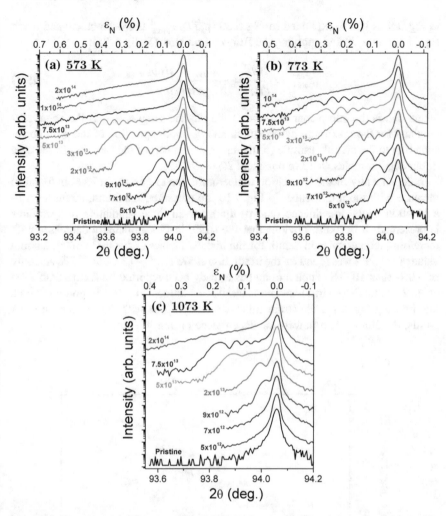

Fig. 9.9 θ-2θ scans recorded in the vicinity of the (400) reflection for pristine and irradiated MgO crystals at increasing Au-ion fluences and at **a** 573 K, **b** 773 K and **c** 1073 K. Labels correspond to the ion fluences (in cm^{-2}). The curves are presented shifted vertically for clarity

Reciprocal space maps (Fig. 9.8) recorded for samples irradiated at the three temperatures and at fluences $\leqslant 7.5 \times 10^{13}$ cm^{-2} do not show any sign of significant diffuse scattering, as has already been observed at RT [27]; this is consistent with a presence of essentially small defects, or, in other words, it confirms the absence of extended defects in this fluence range.

A direct graph-reading of Fig. 9.9 allows the determination, with a good approximation, of the maximum strain level ($\varepsilon_{max}^{total}$) in the irradiated layers [7]; indeed, it is given by the position of the last fringe at the low-angle side of the curves displayed

in Fig. 9.9 as it was explained in Fig. 5.20 (p. 77). $\varepsilon_{max}^{total}$ has been calculated using
Eq. (5.22) on p. 78 and applying the Bragg's Law (Eq. (5.18), p. 75):

$$\varepsilon_{max}^{total} = \frac{\Delta d}{d_{hkl}} = \frac{sin\theta_{min} - sin\theta_{hkl}}{sin\theta_{hkl}} \qquad (9.1)$$

where d_{hkl} is the interplanar distance of the pristine sample, θ_{min} is the value of θ
obtained for the maximum deformation, and θ_{hkl} is the position of the peak corresponding to the pristine sample (or the unirradiated part of the sample). In the case
of our MgO samples we have obtained: $2\theta_{hkl} \sim 94.058$.

Independently of the irradiation temperature, the elastic strain level is found to
increase with the ion fluence. In Fig. 9.10 the measured maximum strain level as
a function of the ion fluence (and of dpa), for all three irradiation temperatures
is presented. It becomes obvious that the strain level significantly decreases with
increasing temperature. In detail, for the highest fluence (7.5×10^{13} cm^{-2}), a strain
value of $\sim 0.59\%$ was found for the irradiation at 573 K, of $\sim 0.32\%$ at 773 K, and only
of $\sim 0.18\%$ at 1073 K. Such a trend has already been reported for Yttria-Stabilized
cubic Zirconia (YSZ) irradiated upon similar conditions [20]. It is important to note
that in a log log scale, the strain follows a straight line with the fluence; in other
words, the strain has a power-law dependence on ion fluence. At 573 and 773 K,

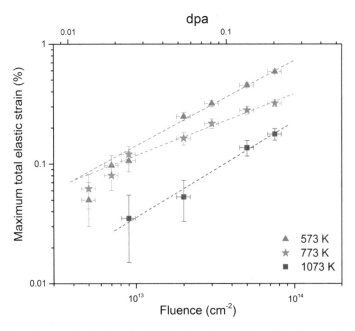

Fig. 9.10 Maximum total elastic strain, derived from the experimental XRD curves displayed in
Fig. 9.9, plotted as a function of both the Au-ion fluence (bottom axis) and the dpa value (top axis)
for the three indicated temperatures. Note that a log-log scale has been used

the linear fitting is not perfect at very low fluence. Nonetheless, it must be noted that even if one discards the first point of the two data sets, the slope of the lines remains the same as for a fit with all data points. A slope of ~0.7 is found for 573 K, it decreases to ~0.5 for 773 K and, surprisingly, it is higher at 1073 K, with a value of ~0.8. This discrepancy will be discussed later. The power-law dependency on dose has also been observed for F centers in neutron irradiated MgO [17]. Van Sambeek et al. [28], using curvature measurements, studied the variation of the strain in MgO crystals irradiated with different ions at the same energy; in particular, for the case of 1 MeV Kr ions at RT, they also found a power-law dependency on the fluence with a value of 0.78. Departing from MgO, such a relationship between elastic strain and ion fluence has also been observed for irradiated Si and Ge crystals analyzed by XRD [26]. In both these cases [26, 28] it was stated that the elastic strain was due to the formation of Frenkel pairs or very small defect clusters. This type of dependency with a power factor inferior to one simply indicates that not all the irradiation-induced defects contribute to the strain because a certain number of them annihilate and/or recombine.

9.2.2 Defect Concentration

The elastic strain and the defect concentration are linked via the following equation [29]:

$$\varepsilon^{def} = \frac{1}{3} \sum_i c_i^{def} \frac{V_i^{rel}}{\Omega}$$

(9.2)

where ε^{def} represents the strain due to the presence of i types of defects of c_i^{def} concentration. V_i^{rel} is the relaxation volume associated to an i-type defect and it is expressed in units of atomic volume Ω. According to the principles of the superposition model [30], when point defects form clusters their displacement fields simply superimpose. This, and hence Eq. (9.2), rests valid as long as the clusters remain small (i.e., approximately below the size of a dislocation loop) and the defects do not interact. Here, both assumptions are valid because this analysis is restricted to the early irradiation stages; in other words, when small defect clusters are formed, their relaxation volume will be proportional to the number of point defects they contain and consequently, the measured elastic strain will follow the same rule. It has to be emphasized that the use of Eq. (9.2) requires the knowledge of the strain due to the defect contribution only and not the experimentally measured strain. Actually, the total measured strain includes the response of the irradiated layer to the stress exerted to it by the bulk, undamaged part of the crystal (usually refers to as a substrate) that inhibits any lateral dimension change, as demonstrated in several cases [7, 24, 28, 31–33]; this is the reason why the left-axis in Fig. 9.10 is labelled 'Maximum *total* elastic strain'. Recently, a simple mechanical model based on linear elasticity theory was proposed to account for this particular strain/stress state in low-energy ion-irradiated materials [7, 8, 33]. This model is valid when the irradiated layer responds elastically to the solicitation presented by the ion irradiation; that is the case at low fluence, i.e., before significant defect clustering and concomitant strain relaxation, as demonstrated for several irradiated materials [7–9], including MgO [27]. Thus it becomes feasible to derive the strain attributed only to defects from the experimentally measured total strain by removing the substrate contribution. This can be achieved when the elastic constants of the material are known and for MgO we used the experimental values of [34] where the elastic compliances were determined in a wide temperature range. These values allowed calculating the proportionality factors between the total and the defect-induced strain at the three temperatures; we found 1.56, 1.61 and 1.69 for 573, 773 and 1073 K, respectively.

The use of Eq. (9.2) also requires the knowledge of the relaxation volumes of the defects. Theoretical estimates of the relaxation volumes of Frenkel pairs in MgO exist, and a value of $\sim$1.8 Ω has been reported for both sublattices [35]. Experimentally, using XRD, a value of (1.0 ± 0.8) Ω was found by Scholz et al. [36], Hickman et al. [37] determined a value of 2.6 Ω, while Henderson et al. derived values between 6 and 8 Ω [38]. Despite the existing differences, for the Frenkel pair relaxation volume in MgO there seems to be a consensus on a value around 2.0 Ω. In order to evaluate more accurately the defect concentration, the defect relaxation volumes were here derived using DFT calculations. These calculations were made by Jean-Paul Crocombette from the CEA-Saclay and they were performed using the VASP code [39] with Projector Augmented Wave formalism and Generalized Gradient Approximation in the Perdew–Becke–Ernzerhofform [40] for the exchange and correlation potential. Mono interstitial and mono-vacancy of cation and oxygen type were considered. The defects were introduced in simulation boxes containing 512 ± 1 atoms and while point defects in MgO may be charged, only neutral defects were considered due to the difficulties related to charged defect relaxation volumes [41]. The obtained values for the relaxation volumes of the different point defects are given in Table 9.1. The Mg Frenkel pair has a larger effect on the lattice than the O one, and interstitials distort more the lattice than vacancies. In order this analysis to remain within the framework of the superposition model every type of defect has to be taken into account. Nevertheless, since both sublattices have the same threshold displacement energy and there is not a big difference in the atomic number of the elements, it can be reasonably considered that the concentrations of Mg and O Frenkel pairs are equal, permitting the use of Eq. (9.2). This assumption leads to an equivalent average value $< V^{rel} >$ of $\sim$2.45 Ω, in reasonable agreement to previously reported values.

Table 9.1 Relaxation volumes of point defects in MgO derived by DFT calculations. Values are given in units of atomic volume (Ω)

	Interstitial	Vacancy	Frenkel pair
Mg	3.65	−0.01	3.64
O	1.15	0.10	1.25

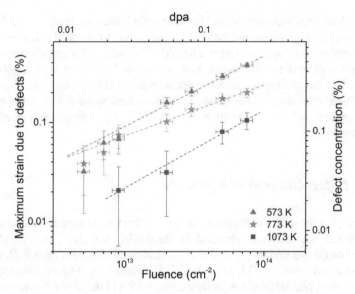

Fig. 9.11 Maximum elastic strain due to defects only (left) and corresponding defect concentrations (right) as a function of both the Au-ion fluence (bottom axis) and the dpa value (top axis) for the three indicated temperatures. Note the log-log scale used

Both ε_{max}^{def} and c_{max}^{def} are plotted in Fig. 9.11. ε_{max}^{def} obviously follows the same trend as the total strain $\varepsilon_{max}^{total}$ since both quantities are proportional. Similarly, the defect concentration increases with the ion fluence, and decreases with increasing temperature; for instance, for the highest fluence 7.5×10^{13} cm^{-2}, c^{def} equals $\sim$0.46% at 573 K, $\sim$0.24% at 773 K and $\sim$0.13% at 1073 K. These values (and those obtained for ε_{max}^{def}) are summarized in Table 9.2. From RT irradiation experiments on MgO [9], the defect concentration before the onset of plastic relaxation was found $\sim$0.53% and this value only slightly decreases (to $\sim$0.46%) when the temperature increases to 573 K. Therefore, defect mobility, and hence defect recovery, does not vary significantly in this temperature range. On the contrary, when temperature increases to 773 K or higher, an efficient dynamic annealing process is observed. Vacancies have been found to be mobile around 873 K in MgO [17] and our results are consistent with this estimate; the observed decrease in the defect concentration is most likely due to

Table 9.2 Values of the maximum total elastic strain ($\varepsilon_{max}^{total}$), maximum strain due to defects (ε_{max}^{def}), defect concentration (c^{def}) and defect generation efficiency for MgO irradiated with Au ions at a fluence of 7.5×10^{13} cm^{-2} at 573, 773 and 1073 K

T (K)	$\varepsilon_{max}^{total}$ (%)	ε_{max}^{def} (%)	c^{def} (%)	Efficiency (%)
573	0.59	0.38	0.46	2.4
773	0.32	0.20	0.24	1.2
1073	0.18	0.11	0.13	0.6

interstitial-vacancy recombination. However, with increasing temperature, interstitial mobility also increases, and their clustering could be enhanced at the expense of their annealing, as being observed in YSZ [20]. This would change their survival rate and might explain the unexpectedly high strain rate increase at 1073 K, though to fully confirm this assumption further work is required. Nonetheless, preliminary ion channeling results tend to indicate the formation of extended defects takes place at lower fluence at 1073 K, compared to 573 and 773 K, suggesting an enhanced defect clustering process.

9.2.3 Defect Generation Efficiency

If we compare the defect concentrations derived from XRD experiments to the estimated defect concentrations obtained by the SRIM code [2, 3] in 'Full Damage Cascade' mode, we find very low defect generation efficiencies (Table 9.2), namely $\sim$2.4, $\sim$1.2 and $\sim$0.6% at 573, 773 and 1073 K, respectively (and $\sim$2.7% at RT from [28]). Previous MD simulations indicated that $\sim$15% of the atoms displaced during a collision cascade induced by a 5 keV oxygen recoil survive [19]. Interstitials are mobile at RT in MgO, and this value of 15% should further decrease with time after the picosecond that the cascade lasts. Nevertheless, this predicted efficiency remains quite high. In their work on 1 MeV Kr MgO irradiation, Van Sambeek et al. [18] calculated this efficiency factor to be 26%, but this value can be hardly compared with ours since it was determined with respect to defect concentrations obtained with the modified Kinchin Pease model [46] (and using V^{rel}=1.8 Ω). We thus performed SRIM calculations (in 'Full Damage Cascade' mode) for 1 MeV Kr in MgO to recalculate their efficiency factor. We found, using our average value $< V^{rel} >= 2.45$ Ω, an efficiency of $\sim$5.5% at RT (for a fluence at strain saturation of 10^{14} cm^{-2}). Again, although it is the same order of magnitude to that obtained in the present work, this value remains high. One reason likely explaining this discrepancy may be related to the electronic energy loss, which is much higher in the present irradiation conditions than in Van Sambeek's experiments (4 keV/nm versus 1.6 keV/nm). Earlier studies pointed out an ionization-induced annealing effect in MgO [47,48,49], and defect recovery has been found to take place even upon X ray irradiation [42]. Van Sambeek et al. performed 1 MeV He irradiation in pre-damaged, Kr irradiated MgO, and they also observed an annealing effect [28, 43]. The present results are supportive to the effects of the electronic excitations on the damage creation processes in irradiated MgO.

9.3 Conclusions

The structural disorder in MgO under 1.2 MeV Au^+ irradiation at 573, 773 and 1073 K has been studied using RBS/C and XRD. RBS/C data reveal the existence of two steps in the damage accumulation process for the three temperatures. The analysis of the XRD reciprocal space maps confirms this result and, furthermore, suggests that the defects formed at high temperature are similar to those previously observed at RT: in the first step point defects are generated, and they grow in the second step forming dislocation loops. However, two particularities have been identified for increasing irradiation temperature: the disorder level decreases, and the second step takes place at higher fluence. Moreover, the location of the damage peak depends on the irradiation temperature: the higher the temperature, the deeper the peak location. We ascribe these results to an enhancement of the defect mobility at higher temperature, which facilitates defect migration and may favor defect annealing.

High resolution XRD has been used to obtain the irradiation-induced elastic strain in MgO during the first step observed with RBS/C. Assuming that at low fluence only point defects or very small defect clusters are formed, the defect concentration could be deduced from the measured total elastic strain. In order to correlate elastic strain and defect concentration, the required Mg and O point defect relaxation volumes have been calculated using DFT. It has been found that, at all temperatures, the defect concentration increases with the ion fluence. However, a dynamic annealing effect is clearly observed at temperatures equal to or higher than 773 K. The defect generation efficiencies have been estimated and have been found to be very small: of the order of 1%. An annealing effect due to electronic energy deposition is suspected to explain these low efficiencies.

References

1. K. Jin, Electronic Energy Loss of Heavy Ions and Its Effects in Ceramics Electronic Energy Loss of Heavy Ions and Its Effects in Ceramics, Ph.D. thesis, University of Tennessee, 2015
2. J.F. Ziegler, J.P. Biersack, U. Littmark, *The Stopping and Range of Ions in Solids* (Pergamon, New York, 1985). http://www.srim.org
3. J.F. Ziegler, *SRIM: The Stopping and Range of Ions in Matter*. http://www.srim.org/
4. S.J. Zinkle, C. Kinoshita, Defect production in ceramics. J. Nucl. Mater. **251**, 200–217 (1997)
5. L. Nowicki, A. Turos, R. Ratajczak, A. Stonert, F. Garrido, Modern analysis of ion channeling data by Monte Carlo simulations. Nucl. Instrum. Methods Phys. Res. Sect. B Beam Interact. Mater. Atoms **240**(1–2), 277–282 (2005)
6. A. Debelle, L. Thomé, A. Boulle, S. Moll, F. Garrido, L. Qasim, P. Rosza, Response of cubic zirconia irradiated with 4-MeV Au ions at high temperature: an X-ray diffraction study. Nucl. Instrum. Methods Phys. Res. Sect. B Beam Interact. Mater. Atoms **277**, 14–17 (2012); Basic research on ionic-covalent materials for nuclear applications
7. A. Debelle, A. Declémy, XRD investigation of the strain/stress state of ion-irradiated crystals. Nucl. Instrum. Methods Phys. Res. Sect. B Beam Interact. Mater. Atoms **268**(9), 1460–1465 (2010)
8. A. Debelle, A. Boulle, F. Garrido, L. Thomé, Strain and stress build-up in He-implanted UO_2 single crystals: an X-ray diffraction study. J. Materi. Sci. **46**(13), 4683–4689 (2011)

9. S. Moll, Y. Zhang, A. Debelle, L. Thomé, J.P. Crocombette, Z. Zihua, J. Jagielski, W.J. Weber, Damage processes in MgO irradiated with medium-energy heavy ions. Acta Mater. **88**, 314–322 (2015)

10. J. Jagielski, L. Thomé, Multi-step damage accumulation in irradiated crystals. Appl. Phys. A **97**(1), 147–155 (2009)

11. A. De Vita, M.J. Gillan, J.S. Lin, M.C. Payne, I. Stich, L.J. Clarke, Defect energetics in oxide materials from first principles. Phys. Rev. Lett. **68**, 3319–3322 (1992)

12. B.P. Uberuaga, R. Smith, A.R. Cleave, G. Henkelman, R.W. Grimes, A.F. Voter, K.E. Sickafus, Dynamical simulations of radiation damage and defect mobility in MgO. Phys. Rev. B Condens. Matter Mater. Phys. **71**, 104102 (2005)

13. C.A. Gilbert, S.D. Kenny, R. Smith, E. Sanville, Ab initio study of point defects in magnesium oxide. Phys. Rev. B Condens. Matter Mater. Phys. **76**, 184103 (2007)

14. J. Mulroue, D.M. Duffy, An Ab initio study of the effect of charge localization on oxygen defect formation and migration energies in magnesium oxide. Proc. R. Soc. Lond. A Math. Phys. Eng. Sci. **467**, 2054–2065 (2011)

15. S. Moll, A. Debelle, L. Thomé, G. Sattonnay, J. Jagielski, F. Garrido, Effect of temperature on the behavior of ion-irradiated cubic zirconia. Nucl. Instrum. Methods Phys. Res. Sect. B Beam Interact. Mater. Atoms **286**, 169–172 (2012). Proceedings of the Sixteenth International Conference on Radiation Effects in Insulators (REI)

16. M. Kilo, R.A. Jackson, G. Borchardt, Computer modelling of ion migration in zirconia. Philos. Mag. **83**(29), 3309–3325 (2003)

17. C. Kinoshita, K. Hayashi, S. Kitajima, Kinetics of point defects in electron irradiated MgO. Nucl. Instrum. Methods Phys. Res. Sect. B Beam Interact. Mater. Atoms **1**(2–3), 209–218 (1984)

18. W.J. Weber, L. Wang, Y. Zhang, W. Jiang, I.-T. Bae, Effects of dynamic recovery on amorphization kinetics in 6H-SiC. Nucl. Instrum. Methods Phys. Res. Sect. B Beam Interact. Mater. Atoms **266**(12–13), 2793–2796 (2008). Proceedings of the Fourteenth International Conference on Radiation Effects in Insulators

19. Y. Zhang, C.M. Wang, M.H. Engelhard, W.J. Weber, Irradiation behavior of SrTiO$_3$ at temperatures close to the critical temperature for amorphization. J. Appl. Phys. **100**(11), 113533 (2006)

20. A. Debelle, J. Channagiri, L. Thomé, B. Décamps, A. Boulle, S. Moll, F. Garrido, M. Behar, J. Jagielski, Comprehensive study of the effect of the irradiation temperature on the behavior of cubic zirconia. J. Appl. Phys. **115**(18), 183504 (2014)

21. I.O. Usov, J.A. Valdez, K.E. Sickafus, Temperature dependence of lattice disorder in Ar-irradiated (1 0 0), (1 1 0) and (1 1 1) MgO single crystals. Nucl. Instrum. Methods Phys. Res. Sect. B Beam Interact. Mater. Atoms **269**(3), 288–291 (2011)

22. J. Channagiri, A. Boulle, A. Debelle, Diffuse X-ray scattering from ion-irradiated materials: a parallel-computing approach. J. Appl. Crystallogr. **48**(1), 252–261 (2015)

23. V.S. Speriosu, Kinematical x-ray diffraction in nonuniform crystalline films: strain and damage distributions in ion-implanted garnets. J. Appl. Phys. **52**, 6094–6103 (1981)

24. S. Rao, B. He, C.R. Houska, X-ray diffraction analysis of concentration and residual stress gradients in nitrogen-implanted niobium and molybdenum. J. Appl. Phys. **69**(12), 8111–8118 (1991)

25. A. Boulle, A. Debelle, Strain-profile determination in ion-implanted single crystals using generalized simulated annealing. J. Appl. Crystallogr. **43**, 1046–1052 (2010)

26. V.S. Speriosu, B.M. Paine, M.A. Nicolet, H.L. Glass, X-ray rocking curve study of Si-implanted GaAs, Si, and Ge. Appl. Phys. Lett. **40**(7), 604–606 (1982)

27. S. Moll, L. Thomé, G. Sattonnay, A. Debelle, F. Garrido, L. Vincent, J. Jagielski, Multistep damage evolution process in cubic zirconia irradiated with MeV ions. J. Appl. Phys. **106**(7), 073509 (2009)

28. A.I. Van Sambeek, R.S. Averback, Cantilever beam stress measurements during ion irradiation, in *Symposium A – Ion-Solid Interactions for Materials Modification and Processing*, vol. 396 of *MRS Proceedings* (1995), pp. 137–142

29. P. Ehrhart, K.H. Robrock, H. Shober, *Physics of Radiation Effects in Crystals, chapter Basic defects in metals* (Elsevier, Amsterdam, Netherlands, 1986)

30. P.H. Dederichs, The theory of diffuse X-ray scattering and its application to the study of point defects and their clusters. J. Phys. F Met. Phys. **3**(2), 471 (1973)

31. J.D. Kamminga, T.H. de Keijser, R. Delhez, E.J. Mittemeijer, On the origin of stress in magnetron sputtered TiN layers. J. Appl. Phys. **88**(11), 6332–6345 (2000)

32. A. Debelle, G. Abadias, A. Michel, C. Jaouen, Stress field in sputtered thin films: Ion irradiation as a tool to induce relaxation and investigate the origin of growth stress. Appl. Phys. Lett. **84**(24), 5034–5036 (2004)

33. A. Debelle, A. Boulle, F. Rakotovao, J. Moeyaert, C. Bachelet, F. Garrido, L. Thomé, Influence of elastic properties on the strain induced by ion irradiation in crystalline materials. J. Phys. D Appl. Phys. **46**(4), 045309 (2013)

34. Y. Sumino, O.L. Anderson, I. Suzuki, Temperature coefficients of elastic constants of single crystal MgO between 80 and 1300 K. Phys. Chem. Miner. **9**(1), 38–47 (1983)

35. M.J.L. Sangster, D.K. Rowell, Calculation of defect energies and volumes in some oxides. Philos. Mag. A **44**, 613–624 (1981)

36. C. Scholz, P. Ehrhart, F-Centers and Oxygen-Interstitials in MgO, in *Symposium A – Beam-Solid Interactions – Fundamentals and Applications*, vol. 279 of *MRS Proceedings* (1992), pp. 427–432

37. B.S. Hickman, D.G. Walker, Growth of magnesium oxide during neutron irradiation. Philos. Mag. **11**, 1101–1108 (1965)

38. B. Henderson, D.H. Bowen, Radiation damage in magnesium oxide. I. dose dependence of reactor damage. J. Phys. C Solid State Phys. **4**(12), 1487–1495 (1971)

39. G. Kresse, D. Joubert, From ultrasoft pseudopotentials to the projector augmented-wave method. Phys. Rev. B Condens. Matter Mater. Phys. **59**, 1758–1775 (1999)

40. J.P. Perdew, K. Burke, M. Ernzerhof, Generalized gradient approximation made simple. Phys. Rev. Lett. **77**, 3865–3868 (1996)

41. F. Bruneval, C. Varvenne, J.P. Crocombette, E. Clouet, Pressure, relaxation volume, and elastic interactions in charged simulation cells. Phys. Rev. B Condens. Matter Mater. Phys. **91**, 024107 (2015)

42. H. Ogiso, S. Nakano, J. Akedo, Abnormal distribution of defects introduced into MgO single crystals by MeV ion implantation. Nucl. Instrum. Methods Phys. Res. Sect. B Beam Interact. Mater. Atoms **206**, 157–161 (2003). 13th International Conference on Ion Beam Modification of Materials

43. A.I. Van Sambeek, *Radiation-enhanced diffusion and defect production during ion irradiation of MgO and Al_2O_3*, Ph.D. thesis, University of Illinois, Urbana-Champaign, USA, 1997

Chapter 10
Ion Beam Induced Luminescence in MgO

In this last chapter of the thesis the ionoluminescence of the MgO samples (similar to those used in Chap. 9) has been studied.

The aim of this chapter is to report on the IL spectra of MgO under different irradiation conditions: at RT, at low temperature ($\sim$100 K), with light ions (1 MeV H^+), and with heavy ions (40 MeV Br^{7+}). The spectra shown in this chapter will be given as a function of the energy (which is essential for the peak analysis), but also, as a function on the wavelength in order to facilitate the comparison with other works, since many of the spectra shown in the literature are presented in this way.

These IL experiments were carried out at the CMAM using the implantation beamline as explained in p. 61. MgO samples were covered with a copper tape to avoid charge effects on the sample surface and to delimit the irradiation area, which was always 7 × 4 mm. The beam was rastered over an area larger than 7 × 4 mm. The temperature of the sample during the irradiations was $\sim$100 K for the low temperature measurements (we will refer to it as LNT for simplicity, although the temperature was higher that 77 K) and $\sim$295 K for the room temperature measurements. The ion beam current was only measured at the beginning and at the end of each measurement. The values of the current, the ion fluxes, and the maximum fluence reached in each irradiation are collected in Table 10.1.

Table 10.1 Current, flux and fluence for the four irradiation conditions

Ion and T	I_i (pnA)	I_f (pnA)	ϕ_i (cm^{-2}·s^{-1})	ϕ_f (cm^{-2}·s^{-1})	Φ_{max} (cm^{-2})
H^+ 100 K	118	118	3.7×10^{11}	3.7×10^{11}	1.1×10^{14}
H^+ 300 K	119	113	5.3×10^{11}	5.0×10^{11}	3.3×10^{14}
Br^{7+} 100 K	4.4	3.8	1.2×10^{10}	1.0×10^{10}	2.7×10^{13}
Br^{7+} 300 K	4.4	5.0	1.3×10^{10}	1.4×10^{10}	3.4×10^{13}

© Springer Nature Switzerland AG 2018

D. Bachiller Perea, *Ion-Irradiation-Induced Damage in Nuclear Materials*, Springer Theses, https://doi.org/10.1007/978-3-030-00407-1_10

10.1   Main Features of the IL Spectrum of MgO

The two first things that we observed when irradiating MgO were: (i) the MgO ionoluminescence is mainly red, and (ii) the IL signal of MgO is very intense (if we compare for example to that of silica). Both effects can be observed in Fig. 10.1 where two pictures taken during MgO irradiation are shown. The picture on the left was taken with the same parameters of the optical camera that were used for the pictures of silica shown in Fig. 6.7 (p. 97), and the difference on the signal intensity is clear. These parameters were modified to take the picture on the right (Fig. 10.1) to better appreciate the color of the emission.

Figure 10.2 shows a MgO spectrum obtained with 1 MeV H^+ at 100 K over the whole measured wavelength range (200–1000 nm). This example is shown to describe the typical MgO IL spectrum expressed as a function of the wavelength (Fig. 10.2a) and as a function of the energy (Fig. 10.2b). Note that the conversion of the intensity explained in p. 65 has been applied for (b).

Table 10.2 summarizes the regions of the MgO IL spectra from Fig. 10.2 and the main features observed in each of them. There are two regions in which an IL signal is observed: 1.33–1.85 eV (930–670 nm) and 2.1–6.0 eV (600–207 nm). Since the intensity of the signals in both regions is extremely different, each region will be studied separately along this chapter and named Region I and Region II, respectively.

Fig. 10.1 Pictures of the luminescence of MgO during proton irradiation taken with different optical camera settings

Table 10.2 Limits and main IL features of the regions into which an IL MgO spectrum can be divided

Region	λ (nm)	E (eV)	IL signal
–	1.24–1.33	1000–930	No IL signal
Region I	1.33–1.85	930–670	Very intense IL emissions
–	1.85–2.07	670–600	No IL signal
Region II	2.07–6.00	600–207	Very weak IL emissions

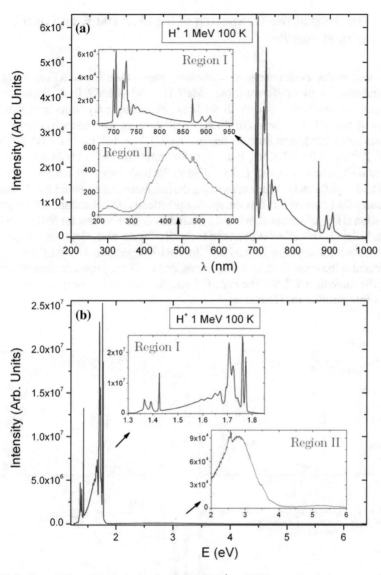

Fig. 10.2 IL spectrum of MgO obtained with 1 MeV H$^+$ at 100 K over the whole measured range as a function of: **a** the wavelength and **b** the energy

10.2   Analysis of the IL Spectra of MgO at 100 K and at RT with H and Br

In this section the spectra obtained at low temperature (100 K) and RT are compared and analyzed for two different ions: 1 MeV H^+ and 40 MeV Br^{7+}. Although the IL signal was monitored during all the irradiation time, only some representative spectra at two (H^+) or three (Br^{7+}) different fluences are shown in this section. The maximum irradiation fluences reached were 1.1×10^{14} cm^{-2} for H^+ at 100 K, 3.3×10^{14} cm^{-2} for H^+ at RT, and $\sim 3 \times 10^{13}$ cm^{-2} for Br^{7+} at both temperatures.

Figure 10.3 shows Region I (1.33–1.85 eV) for both ions, both temperatures, and at low and high fluences. When comparing the influence of the ions on the IL spectra, the main effect observed is that the spectra practically do not change with light ion irradiation (H^+), while the IL intensity decreases extremely fast for SHI irradiation (40 MeV Br^{7+}). This effect is observed for both temperatures. However, the position of the peaks is the same for H^+ and Br^{7+} for each temperature, i.e., at LNT the peaks are located at the same position for both ions, and at RT the peaks (some of which not being the same than at LNT, see Fig. 10.5 and discussion hereafter) are also located at the same position for H^+ and Br^{7+}.

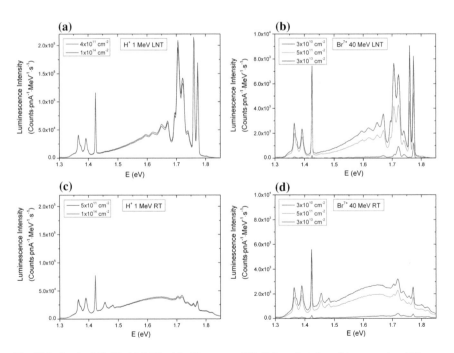

Fig. 10.3   Region I (1.33–1.85 eV) of the IL spectra of MgO (as a function of the energy) at different irradiation fluences obtained with: **a** 1 MeV H^+ at 100 K, **b** 40 MeV Br^{7+} at 100 K, **c** 1 MeV H^+ at RT, and **d** 40 MeV Br^{7+} at RT

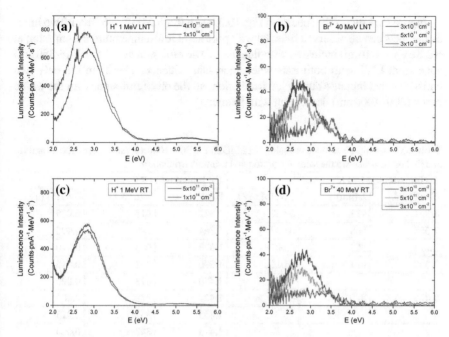

Fig. 10.4 Region II (2.1–6.0 eV) of the IL spectra of MgO (as a function of the energy) at different irradiation fluences obtained with: **a** 1 MeV H$^+$ at 100 K, **b** 40 MeV Br^{7+} at 100 K, **c** 1 MeV H$^+$ at RT, and **d** 40 MeV Br^{7+} at RT

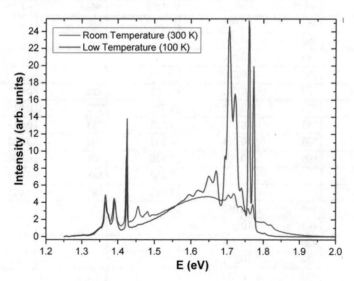

Fig. 10.5 Comparison of an IL spectrum of MgO at RT (red) to a 100 K spectrum (blue). Both spectra were measured with 1 MeV protons

Figure 10.4 shows the spectra of Fig. 10.3 but in Region II (2.1–6.0 eV). Two interesting facts can be observed here: (i) with protons at low temperature the emission at ~2.8 eV (~440 nm) increases with fluence; (ii) an emission at ~3.55 eV (~350 nm) appears at LNT with both ions when increasing fluence. These are the only two emissions that increase during ion irradiation, all the other emissions over the whole range (200–1000 nm) decreasing with fluence.

Table 10.3 Ionoluminescence emissions of MgO in the 200–1000 nm range at (a) 100 K, and (b) 300 K. The IL peaks are the same for proton and bromine irradiation

(a) Emissions at 100 K			(b) Emissions at RT		
E (eV)	λ (nm)	FWHM (eV)	E (eV)	λ (nm)	FWHM (eV)
1.351	918	0.027	1.352	917	0.025
1.365	908	0.007	1.364	909	0.008
1.373	903	0.007	1.373	903	0.010
1.392	891	0.011	1.390	892	0.013
			1.420	873	0.020
1.425	870	0.003	1.422	872	0.004
			1.455	852	0.013
			1.473	842	0.009
			1.481	837	0.008
1.593	778	0.026			
~1.60	~777	~0.23	~1.63	~761	~0.26
1.622	764	0.025			
1.650	752	0.024			
1.670	743	0.014			
1.693	732	0.007			
1.706	727	0.011	1.703	728	0.012
1.722	720	0.013	1.718	722	0.013
1.740	713	0.014	1.738	713	0.013
1.760	705	0.006	1.757	706	0.009
1.773	699	0.006	1.770	701	0.008
			1.803	688	0.015
			1.820	681	0.013
2.491	498	0.040			
2.517	493	0.017			
2.539	489	0.023			
2.564	484	0.017			
2.588	479	0.017			
2.807	442	0.962	2.833	438	0.844
3.569	347	0.217	3.558	349	0.323
5.212	238	1.240			

Figure 10.5 compares an IL MgO spectrum obtained with protons at 100 K to one at RT (from Fig. 10.3a, c). Some peaks are found at both temperatures, but other peaks are found only at LNT or only at RT.

Since the positions of the peaks do not change with the ions but they do change with the temperature, the analysis of the IL emissions has been done for protons at 100 K and for protons at RT. The analyses have been done by fitting the spectra with Gaussian curves using the program Fityk. All the results of the fits of the spectra at 100 K and RT are summarized in Table 10.3.

Figure 10.6 shows the fit of an IL MgO spectrum obtained with 1 MeV H$^+$ at 100 K in the 1.33–1.85 eV range (Region I, 930–670 nm). This spectrum consists of a broad emission peaking at ∼1.6 eV (∼775 nm) and of 15 sharp emissions (five in the range 1.35–1.43 eV and ten in the range 1.6–1.8 eV).

Figure 10.7 shows the spectrum of Fig. 10.6 but in the 2.1–6.0 eV range (Region II, 600–207 nm). Two broad bands are observed at ∼2.8 eV (∼440 nm) and ∼5.2 eV (∼240 nm). A small band at ∼3.6 eV (∼350 nm) appears at high fluence. Five sharp emissions are observed in the 2.5–2.6 eV region (∼500–470 nm). The two emissions marked with an asterisk (2.52 and 2.54 eV) could be, each of them, formed by two narrow emissions at 2.514 and 2.524 eV for the 2.52 eV peak, and 2.535 and 2.543 eV for the 2.54 eV peak. But these values are too close to be well separated since the resolution of the spectrometer is ∼2 nm.

The same two regions have been analyzed for the RT spectrum. The analysis of Region I is shown in Fig. 10.8. The same broad emission at ∼1.6 eV is observed. Most of the peaks observed at 100 K at 1.35–1.43 and at 1.7–1.8 eV also appear at RT. However, the four peaks present at low temperature at 1.59–1.67 eV do not

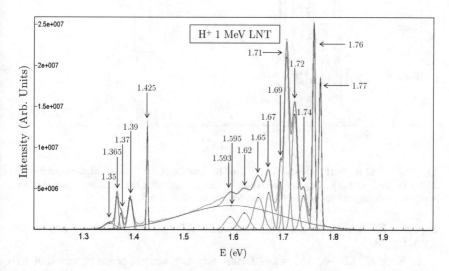

Fig. 10.6 Fit of the 1.33–1.85 eV region (Region I) of the IL spectrum of MgO obtained with 1 MeV H$^+$ at 100 K. The green line is the experimental spectrum, the black line is the simulated spectrum, and the red lines are the Gaussian functions used for the fit

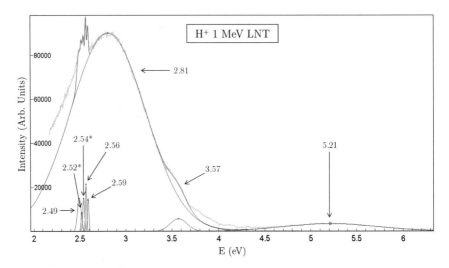

Fig. 10.7 Fit of the 2.1–6.0 eV region (Region II) of the IL spectrum of MgO obtained with 1 MeV H$^+$ at 100 K. The green line is the experimental spectrum, the black line is the simulated spectrum, and the red lines are the Gaussian functions used for the fit

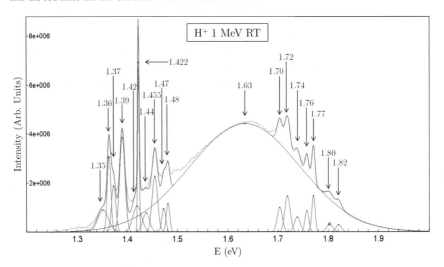

Fig. 10.8 Fit of Region I (1.33–1.85 eV, 930–670 nm) of the IL spectrum of MgO obtained with 1 MeV H$^+$ at RT (300 K). The green line is the experimental spectrum, the black line is the simulated spectrum, and the red lines are the Gaussian functions used for the fit

appear at RT. Seven new peaks are observed at RT: five at 1.42–1.48 eV and two at 1.80–1.82 eV.

In Region II (2.1–6.0 eV, 600–207 nm) only two emissions are observed at RT (Fig. 10.9): the broad emission at ∼2.8 eV, and the small peak at ∼3.6 eV. The narrow emissions around 2.5 eV and the band at 5.2 eV observed at low temperature are not visible at RT.

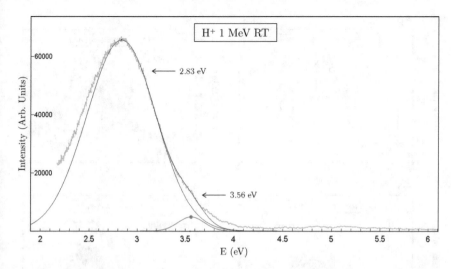

Fig. 10.9 Fit of Region II (2.1–6.0 eV, 600–207 nm) of the IL spectrum of MgO obtained with 1 MeV H⁺ at RT (300 K). The green line is the experimental spectrum, the black line is the simulated spectrum, and the red lines are the Gaussian functions used for the fit

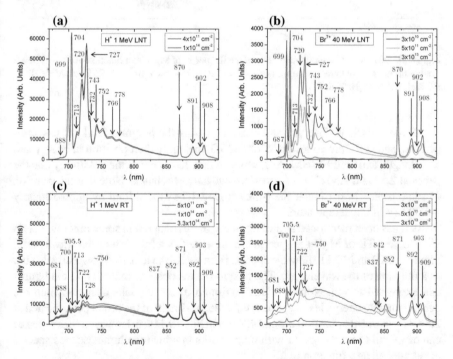

Fig. 10.10 670–930 nm region (Region I) of the IL spectra of MgO (as a function of λ) at different irradiation fluences obtained with: **a** 1 MeV H⁺ at 100 K, **b** 40 MeV Br⁷⁺ at 100 K, **c** 1 MeV H⁺ at RT, and **d** 40 MeV Br⁷⁺ at RT

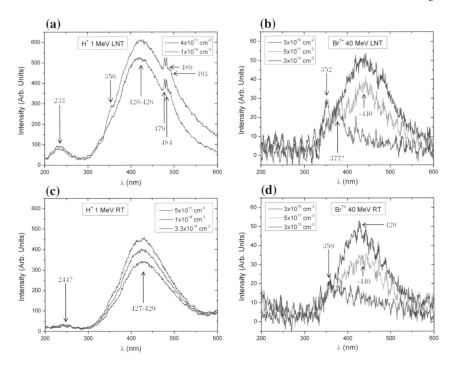

Fig. 10.11 207–600 nm region (Region II) of the IL spectra of MgO (as a function of λ) at different irradiation fluences obtained with: **a** 1 MeV H$^+$ at 100 K, **b** 40 MeV Br^{7+} at 100 K, **c** 1 MeV H$^+$ at RT, and **d** 40 MeV Br^{7+} at RT

All the emissions observed and analyzed (using the program Fityk) at low temperature and at RT are collected in Table 10.3, including their position in energy and wavelength, and their FWHM. As it was said in p. 157, it exists the possibility that the peaks at 2.52 and 2.54 eV are formed by two narrower bands. Note that the position of some peaks (in particular the peaks around 1.7 eV) is shifted by approximately +0.003 eV at low temperature.

As it has been mentioned at the beginning of this chapter, in some other works on CL, PL, and TL of MgO the spectra are presented as a function of the wavelength. Figures 10.10 and 10.11 show the spectra from Figs. 10.3 and 10.4 (respectively) but as a function of the wavelength. The position of the peaks is indicated in the figures and expressed in nm. Note that the maximum of some peaks changes when the conversion into energy is not applied; e.g., the maximum of the broad peak located at 2.8 eV (∼440 nm) appears at around 430 nm in Fig. 10.11a, c. However, these spectra can be useful to be compared with other works in which the luminescence spectra are expressed as a function of λ.

10.3 Kinetics of the Main IL Emissions

In this section the evolution of the intensity of the main IL emissions as a function of the irradiation fluence (the kinetics of the IL emissions) is studied. Only the most intense bands are studied: the sharpest bands between 700 nm and 908 nm, the broad band at $\sim$760 nm, the broad band at $\sim$440 nm, and the small band at $\sim$352 nm. Note that the evolution of the broad band at $\sim$1.6 eV ($\sim$760–770 nm) is studied by taking the intensity of the 790 nm channel, since the maximum of the emission overlaps with other bands at low temperature.

Figure 10.12 shows the kinetics of the main bands between 700 and 908 nm when irradiating with 1 MeV protons at 100–300 K. The intensity of the peaks slightly decreases with fluence, but no drastic changes are observed. The small peaks in the curves correspond to abrupt variations of the beam current intensity.

The kinetics of the main bands between 700 and 908 nm when irradiating with 40 MeV Br^{7+} ions at 100 and 300 K are presented in Fig. 10.13. A very fast decrease of the IL intensity is observed at the very beginning of the irradiation, i.e., up to a fluence of $\sim$2.5 $\times$ 10^{12} cm^{-2}; above this fluence, the intensity continues decreasing but a much lower rate.

The evolution of the 2.8 eV (440 nm) and 3.6 eV (352 nm) bands is studied in different graphs since their intensity is much lower than that of the other emissions.

The behavior of the emission at 2.8 eV (440 nm) does not always follow the same tendency. Figure 10.14 shows the kinetics of the 2.8 eV band for Br ions at both temperatures. In the case of Br irradiations the behavior is similar to that observed for the other emissions: the intensity decreases fast at the beginning at slowly at fluences higher than $\sim$2.5 $\times$ 10^{12} cm^{-2}.

On the contrary, the intensity of this band is observed to increase when irradiating with protons at low temperature (Fig. 10.15). Furthermore, at the very beginning of the irradiation a fast increase of the intensity is observed. This increase cannot be due to variations in the current or other external factors because it is not observed for the other emissions and because the current was constant during this irradiation (Table 10.1). With protons at RT we have not observed this type of behavior, but since the beam current decreased during the irradiation (see Table 10.1), we cannot confirm that this effect does not happen at RT.

In the case of the 3.6 eV (352 nm) band, its kinetics is studied only for Br at LNT (Fig. 10.16), since for the other three cases the signal completely overlaps with the 2.8 eV emission, and thus, the evolution of the yield at this channel would not reflect the behavior of the 3.6 band (see Fig. 10.4). In Fig. 10.16 it can be seen that right at the beginning of the irradiation the intensity decreases (which is due to the decrease of the 2.8 eV emission) but then it increases during the whole irradiation time. This band was also observed to increase with protons at LNT (Fig. 10.4), whereas at RT it practically does not change with fluence (although it is present in the spectrum).

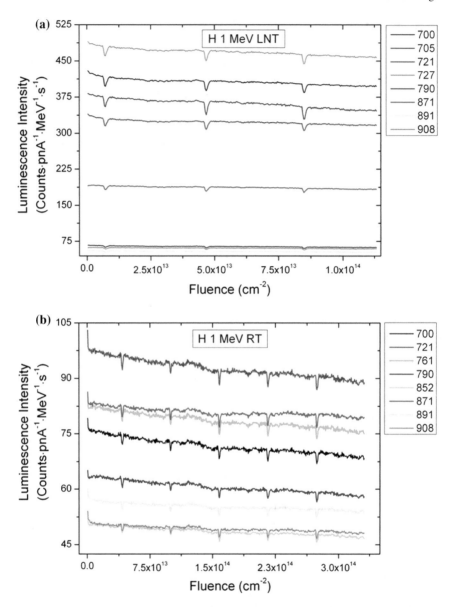

Fig. 10.12 Kinetics of the main IL bands of MgO between 700 nm and 908 nm when irradiating with 1 MeV protons at **a** 100 K and **b** 300 K. Note the different scales for the fluence in (**a**) and (**b**)

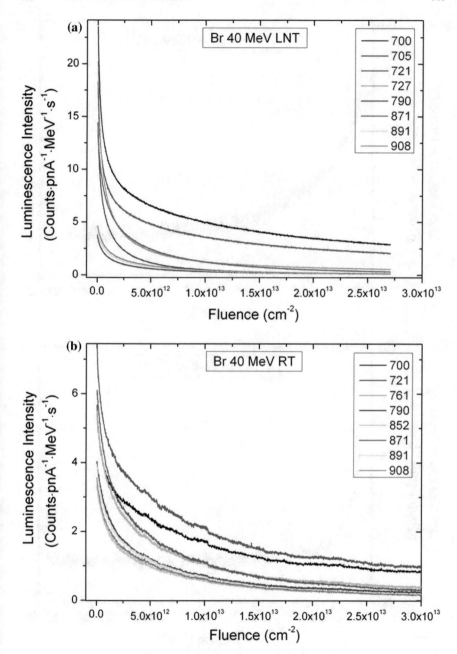

Fig. 10.13 Kinetics of the main IL bands of MgO between 700 nm and 908 nm when irradiating with 40 MeV Br^{7+} ions at **a** 100 K and **b** 300 K

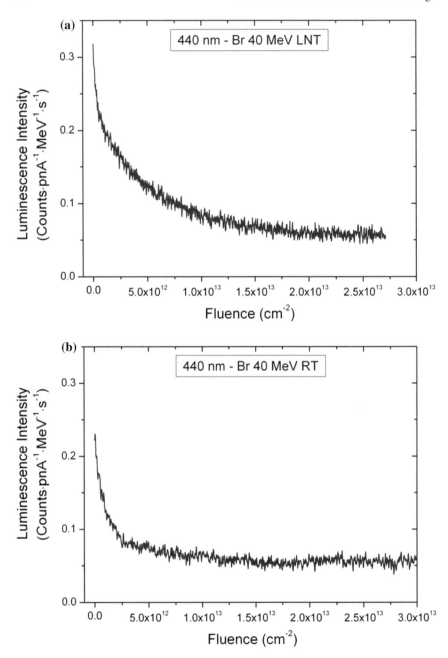

Fig. 10.14 Evolution of the intensity of the 2.8 eV (440 nm) band with fluence for **a** 40 MeV Br^{7+} at 100 K, and **b** 40 MeV Br^{7+} at RT

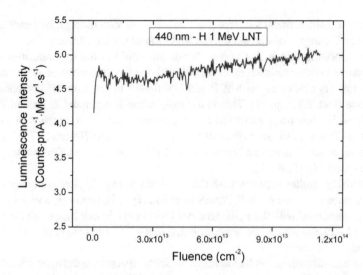

Fig. 10.15   Evolution of the 2.8 eV (440 nm) band with fluence for 1 MeV H$^+$ at 100 K

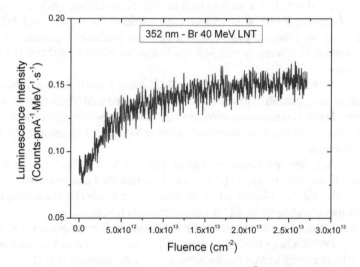

Fig. 10.16   Evolution of the 3.6 eV (352 nm) band for 40 MeV Br^{7+} at 100 K

10.4   Discussion

Several IL bands have been observed and reported here for MgO single crystals. Two different behaviors of the emissions have been observed, which suggests a different origin of certain bands. Previous studies using PL, TL, CL and ESR [1–4] have concluded that the luminescence emissions of MgO arise from hole recombination with trace transition metal ion impurities. In our IL spectra most of the bands are due to the metallic impurities present in the MgO samples (Sect. 2.2.2, p. 25).

One surprising fact is that several bands attributed to the Cr impurities can be well observed with ionoluminescence, although the Cr concentration in our samples is very low (not detected with ICP and practically in the limit of detection with PIXE, see Sect. 2.2.2, p. 25). This is the case of the broad band at around 750 nm (1.6 eV) which has been attributed to the presence of Cr^{3+} impurities in rhombic centers [5–7], and of the sharp lines at 699 nm (1.77 eV) and 704 nm (1.76 eV) which correspond to the R-lines and N-lines of the Cr^{3+} impurities in cubic and tetragonal sites, respectively [5, 8–10].

Something similar happens with the V impurity: the supplier company did not find any trace of V using the ICP/AES technique (p. 26); however, a small quantity of V was measured with the PIXE analysis (~83 ppm). In our IL spectra the band at 870 nm (1.425 eV) which has been attributed to the V^{2+} impurities [6, 11] is clearly observed.

Other emissions that can be identified in our IL spectra are the ones related to the Fe^{3+} impurities: the sharp bands peaking at 705, 720, 727, and 752 nm [6, 11].

Some of the IL emissions are observed at both temperatures, while there are others that only appear at one of the temperatures (either at 100 or at 300 K). While the intensity does not practically change with light-ion irradiation (protons), it drastically changes at low fluences with SHI irradiation (in this case, 40 MeV Br) at both temperatures.

The bands at 2.8 eV (440 nm) and 3.6 eV (352 nm) seem to have a different origin because they exhibit a different behavior than the emissions due to metallic impurities. These two bands are probably due to lattice defects that appear in MgO when irradiating the material since their intensity increases with fluence under some irradiation conditions.

In some scientific publications, a band due to lattice defects has been found between 400 and 500 nm [6, 11]. This could explain the band observed at 440 nm (2.8 eV) which have a different behavior under proton irradiation at low temperature (its intensity increases, while all the other emissions decrease).

The small bands appearing under Br irradiation at low temperature at 352 nm (3.6 eV) and ~380 nm (3.6 eV) (see Figs. 10.11d and 10.16) could be related to F^+ and F_2 defects arising in MgO as a consequence of the irradiation [12].

Since the ionoluminescence of MgO has not practically been studied yet, these results are still difficult to understand and to interpret, but they provide a good preliminary database for future works. The observation of certain bands in our spectra prove the power of the IL technique to detect impurities in such small concentrations that sometimes they cannot even be detected with other techniques.

References

1. W.W. Duley, M. Rosatzin, The orange luminescence band in MgO crystals. J. Phys. Chem. Solids **46**(2), 165–170 (1985)
2. W.C. Las, T.G. Stoebe, Thermoluminescent mechanisms involving transition metal ion impurities and V-type centres in MgO crystals exposed to ultraviolet radiation. J. Mater. Sci. **17**(9), 2585–2593 (1982)
3. W.A. Sibley, J.L. Kolopus, W.C. Mallard. A study of the effect of deformation on the ESR, luminescence, and absorption of MgO single crystals. Physica status solidi (b) **31**(1), 223–231 (1969)
4. S. Datta, I. Boswarva, D. Holt, SEM cathodoluminescence studies of heat-treated MgO crystals. J. de Physique Colloques **41**(C6), 522–525 (1980)
5. M.O. Henry, J.P. Larkin, G.F. Imbusch, Nature of the broadband luminescence center in MgO:Cr^{3+}. Phys. Rev. B: Cond. Matter Mater. Phys. **13**, 1893–1902 (1976)
6. CSIRO Luminescence Database. http://www.csiro.au/luminescence/
7. E. Shablonin, Processes of structural defect creation in pure and doped MgO and NaCl single crystals under condition of low or super high density of electronic excitations. Ph.D. thesis, University of Tartu, Estonia, 2013
8. G.F. Imbusch, A.L. Schawlow, A.D. May, S. Sugano, Fluorescence of MgO: Cr^{3+} Ions in noncubic sites. Phys. Rev. **140**(3A), 830–838 (1965)
9. V. Skvortsova, L. Trinkler, The optical properties of magnesium oxide containing transition metal ions and defects produced by fast neutron irradiation, *Advances in Sensors, Signals and Materials* (2010)
10. V. Skvortsova, L. Trinkler, Luminescence of impyrity and radiation defects in magnesium oxide irradiated by fast neutrons. Phys. Procedia **2**(2), 567–570 (2009). The 2008 International Conference on Luminescence and Optical Spectroscopy of Condensed Matter
11. C.M. MacRae, N.C. Wilson, Luminescence database I-Minerals and materials. Microsc. Microanal. **14**, 184–204 (2008)
12. D. Kadri, A. Mokeddem, S. Hamzaoui, Intrinsic defects in UV-irradiated MgO single crystal detected by thermoluminescence. J. Appl. Sci. **5**(8), 1345–1349 (2005)

Chapter 11
Conclusions and Prospects for the Future

A huge amount of new results have been obtained during this thesis. Probably one of the most valuable results obtained in this work is the demonstration of the capability and the power of the ionoluminescence technique (IL, which is not a widely known IBA technique) to study in situ the level and evolution of the microscopic and macroscopic damage in materials. During this thesis the development of the IL technique has been implemented in new beamlines (such as the implantation beamline at CMAM); or even in other accelerator facilities, such as the JANNuS-Saclay platform (see Appendix A). This development of the technique has allowed for carrying out new IL experiments under various irradiation conditions, and it will allow to use ionoluminescence as a common tool to characterize materials.

Besides, other important conclusions have been obtained with this thesis and new possibilities are open for future works in the field of ion-irradiation in materials, in particular for materials with nuclear applications. The main results of each part of this thesis and the future work that should be done in this field are summarized hereafter.

11.1 Ion Beam Induced Luminescence in Amorphous Silica

A novel set of data on IL spectra and their kinetics in silica has been obtained. The luminescence induced in three types of silica (with different OH content) by light and heavy ion irradiations ($2\,\mathrm{MeV}$ H^+, $4\,\mathrm{MeV}$ He^+, $4\,\mathrm{MeV}$ C^+, $28.55\,\mathrm{MeV}$ Si^{6+}, $18\,\mathrm{MeV}$ Br^{4+}, and $28\,\mathrm{MeV}$ Br^{6+}) has been comparatively investigated.

Two main IL emissions have been observed at $1.9\,\mathrm{eV}$ ($750\,\mathrm{nm}$, red band) and $2.7\,\mathrm{eV}$ ($460\,\mathrm{nm}$, blue band) corresponding to the recombination of STEs at NBOH and ODC centers, respectively. IL experiments at $100\,\mathrm{K}$ showed that the intrinsic recombination of STEs produces a band at around $2.2\,\mathrm{eV}$ that cannot be observed at RT. It has also been proven that the nuclear contribution to the IL spectra is negligible.

© Springer Nature Switzerland AG 2018
D. Bachiller Perea, *Ion-Irradiation-Induced Damage in Nuclear Materials*,
Springer Theses, https://doi.org/10.1007/978-3-030-00407-1_11

It has been observed that increasing the OH content strongly enhances the yield of the red emission in comparison to that of the blue one. This is in accordance with the operation of an extrinsic channel for the red emission, involving a fast initial scission of the $Si - O - H$ bonds by electronic excitation. These results extend those obtained by Nagata et al.. [1] for H irradiations and help to elucidate the role of those extrinsic bonds on the light emission and color center production in irradiated silica.

It has also been observed that the IL kinetics strongly depends on the stopping power of the incident ions finding a similar dependence on this parameter as the one observed for the structural disorder measured by spectroscopic methods [2]. The kinetic curves for the emission yields as a function of fluence consist of a fast growing stage up to a maximum followed by a decreasing stage down to a final steady value. The fluence corresponding to the maximum yield decreases steadily with increasing stopping power, extending from around $2 \times 10^{15}\,cm^{-2}$ (for 2 MeV protons having $S_e = 0.027\,keV/nm$ in silica samples containing $\sim$600 ppm of OH groups) down to around $5 \times 10^{11}\,cm^{-2}$ for stopping power $S_e = 6\,keV/nm$.

A phenomenological model based on STE generation by ion-beam irradiation, their hopping motion through the silica network, and their recombination at NBOH and ODCII centers to give the red and blue emissions, respectively, has been proposed. At room temperature, the intrinsic STE recombination does not contribute to the light emission, but it does determine the diffusion length traveled during their lifetime and so the IL rate. According to the model, the decreasing stage of the kinetic curves is associated with the change in the diffusion length of the STEs as a consequence of the structural damage caused by the irradiation. A mathematical formulation of the physical model provides a reasonable fit of the two-stage kinetics data for the blue and red bands in silica. Although the IL mechanism described in this work is, in principle, independent of the mechanisms for color center production, it provides interesting information on the synergy between microscopic damage (coloring) and macroscopic damage (structural modifications and compaction).

In summary, the novel results obtained by comparing the IL behavior under light and heavy ion irradiation offer a useful tool to investigate structural damage and compaction in silica samples.

11.2 Ion-Irradiation Damage in MgO

The ion-irradiation damage (in particular, the structural disorder) in MgO single crystals under 1.2 MeV Au^+ irradiation at three temperatures (573, 773 and 1073 K) has been studied using RBS/C and XRD.

RBS/C data reveal the existence of two steps in the damage accumulation process for the three temperatures. The analysis of XRD reciprocal space maps confirms this result and, furthermore, suggests that the defects formed at high temperature are similar to those previously observed at RT: in a first step, point defects are generated, and they grow in a second step forming dislocation loops. However, two particularities have been identified when increasing irradiation temperature: the disorder level decreases, and the second step starts at higher fluence. Moreover, the location of the damage peak depends on the irradiation temperature: the higher the temperature, the deeper the peak location. We ascribe these results to an enhancement of the defect mobility at higher temperature, which facilitates defect migration and may favor defect annealing.

High-resolution XRD has been used to obtain the irradiation-induced elastic strain. Assuming that at low fluence only point defects or very small defect clusters are formed, the defect concentration could be deduced from the measured total elastic strain. In order to correlate elastic strain and defect concentration, the required Mg and O point-defect relaxation volumes have been calculated using DFT. It has been found that at all temperatures, the defect concentration increases with the ion fluence. However, a dynamic annealing effect has been clearly observed at temperatures equal to or higher than 773 K. The defect generation efficiencies have been estimated and were found very small, on the order of 1%; an annealing effect due to electronic energy deposition is suspected to explain these low efficiencies.

The ionoluminescence of MgO single crystals at 100–300 K with light (1 MeV H^+) and swift-heavy ions (40 MeV Br^{7+}) has been studied. Before this thesis the IL of MgO had not been studied. The results presented here offer a new and preliminary database to characterize the main IL emissions of MgO at low temperature and room temperature. The IL bands observed have been attributed to metallic impurities of the samples or to irradiation-induced defects in MgO.

11.3 Prospects for the Future

These results may lead to other studies that could be done in the future, some of which are currently being carried out.

In 2013 a collaboration was established in the framework of this thesis with the JANNuS-Saclay laboratory (see Appendix A). The IBIL experimental system was installed in the triple beam irradiation chamber [3]. A series of experiments were carried out in 2015 to study the IL of silica under double beam irradiation and under sequential irradiations. The results of these experiments are currently being analyzed. The installation of the IBIL setup at this unique facility will allow to study the ionoluminescence of materials under triple beam irradiation and to compare with sequential irradiations to study the possible synergistic effects.

The new physical model and its mathematical formulation proposed here to explain the IL behavior in silica should be applied to determine the microscopic (color center creation) and macroscopic (amorphization) cross-sections in silica. Some preliminary fits have been done here, however more work need to be done by fitting more experimental data of the IL kinetics and studying the dependence of the microscopic and macroscopic cross-sections on the stopping power of the ions.

Regarding the ion-irradiation damage with 1.2 MeV Au in MgO, the same experiments but at low temperature (80 K) are being carried out to compare the irradiation effects in a larger range of temperature. The results obtained in this thesis will be used to implement a kinetic Monte Carlo simulation code of damage evolution with fluence and temperature in collaboration with other research groups.

The ionoluminescence results obtained here for MgO showed that it is not easy to identify the origin of each IL emission, and therefore, further measurements need to be carried out to understand the origin of the bands. Comparative IL experiments could be done with MgO samples doped with different amounts of impurities to determine the type of impurity that causes each IL emission.

The same techniques and models that have been used in this thesis can be applied to study other amorphous or crystalline materials.

The progress in this field of research is essential to understand the damage processes in materials which will allow us to move forward to new systems of producing energy in a clean, efficient and sustainable way.

References

1. S. Nagata, S. Yamamoto, A. Inouye, B. Tsuchiya, K. Toh, T. Shikama, Luminescence characteristics and defect formation in silica glasses under h and he ion irradiation. J. Nucl. Mater. **367–370**(B), 1009–1013 (2007). Proceedings of the Twelfth International Conference on Fusion Reactor Materials (ICFRM-12)
2. K. Awazu, S. Ishii, K. Shima, S. Roorda, J.L. Brebner, Structure of latent tracks created by swift heavy-ion bombardment of amorphous sio_2. Phys. Rev. B Condens. Matter Mater. Phys. **62**, 3689–3698 (2000)
3. L. Beck, Y. Serruys, S. Miro, P. Trocellier, E. Bordas, F. Leprêtre, D. Brimbal, T. Loussouarn, H. Martin, S. Vaubaillon, S. Pellegrino, D. Bachiller-Perea, Ion irradiation and radiation effect characterization at the JANNUS-Saclay triple beam facility. J. Mater. Res. **30**(9), 1183–1194 (2015)

Appendix A
Implantation of the Ionoluminescence
Technique at the JANNuS-Saclay Laboratory

The JANNuS-Saclay platform is a unique facility in Europe which couples three ion accelerators allowing to perform irradiations with two or three ion beams in confluence (double or triple beam irradiations) [1–3].

The ionoluminescence setup presented in Sect. 5.1 was installed at the triple beam chamber (see Fig. A.1). Some tests measuring the IL of silica were made in 2013 to

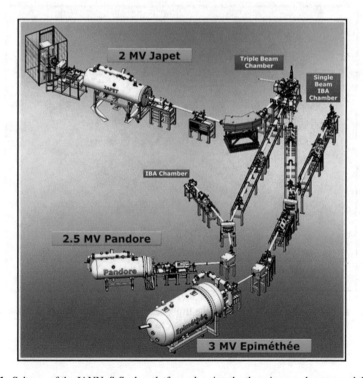

Fig. A.1 Schema of the JANNuS-Saclay platform showing the three ion accelerators and the three experimental chambers. Picture obtained from [3]

© Springer Nature Switzerland AG 2018   173
D. Bachiller Perea, *Ion-Irradiation-Induced Damage in Nuclear Materials*,
Springer Theses, https://doi.org/10.1007/978-3-030-00407-1

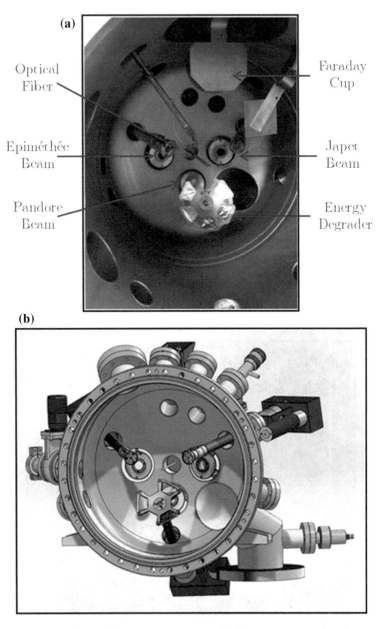

Fig. A.2 **a** Picture showing the inside of the triple beam chamber at JANNuS. **b** Schema of the chamber (obtained from [3]), the optical fiber is placed perpendicular to the sample

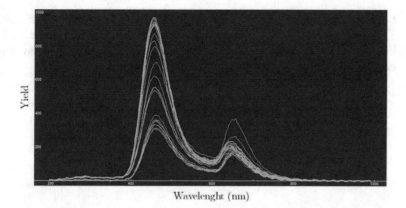

Wavelenght (nm)

Fig. A.3 Example of some IL spectra (at different fluences) of KU1 silica measured during the tests of the experimental setup at JANNuS. These spectra were recorded during double beam irradiation with He and Ti ions

verify that the system worked well. In 2015, a series of experiments were carried out to compare the IL in silica with double irradiation to the IL produced by sequential irradiation with the same ion beams. Ion beams of H, He, and Si were combined. The results obtained are under analysis. In the future, other works comparing the IL produced in different materials by triple beam irradiation to the IL produced by sequential irradiations should be carried out to study the synergistic effects of multiple beam irradiations.

Figure A.2 shows the installation of the optical fiber in the triple beam chamber to collect the light emitted during ion irradiation. The optical fiber was connected to the QE65000 spectrometer as described in Sect. 5.1.

Figure A.3 shows some IL spectra of a KU1 silica sample obtained during irradiation with a double beam of He and Ti ions. These spectra were recorded during the tests of the experimental setup, each spectrum corresponds to a different irradiation fluence.

References

1. Y. Serruys, P. Trocellier, S. Miro, E. Bordas, M.O. Ruault, O. Kaitasov, S. Henry, O. Leseigneur, Th Bonnaillie, S. Pellegrino, S. Vaubaillon, D. Uriot, JANNUS: a multi-irradiation platform for experimental validation at the scale of the atomistic modelling. J. Nucl. Mater. **386–388**, 967–970 (2009). Proceedings of the Thirteenth International Conference on Fusion Reactor Materials
2. S. Pellegrino, P. Trocellier, S. Miro, Y. Serruys, É. Bordas, H. Martin, N. Chaâbane, S. Vaubaillon, J.P. Gallien, L. Beck, The JANNUS Saclay facility: a new platform for materials irradiation, implantation and ion beam analysis. Nucl. Instrum. Methods Phys. Res. Sect. B Beam Interact. Mater. Atoms **273**, 213–217 (2012). 20th International Conference on Ion Beam Analysis

3. L. Beck, Y. Serruys, S. Miro, P. Trocellier, E. Bordas, F. Leprêtre, D. Brimbal, T. Loussouarn, H. Martin, S. Vaubaillon, S. Pellegrino, D. Bachiller-Perea, Ion irradiation and radiation effect characterization at the JANNUS-Saclay triple beam facility. J. Mater. Res. **30**(9), 1183–1194 (2015)

Appendix B
Example of an Input and an Output File from SRIM

The following figures show the input parameters used to obtain the Stopping/Range Tables with the program SRIM, and the corresponding output text file. In this example the stopping powers and the ion range have been calculated for gold ions between 1.0 and 1.4 MeV impinging in amorphous silica.

© Springer Nature Switzerland AG 2018
D. Bachiller Perea, *Ion-Irradiation-Induced Damage in Nuclear Materials*,
Springer Theses, https://doi.org/10.1007/978-3-030-00407-1

```
==================================================================
            Calculation using SRIM-2006
            SRIM version ---> SRIM-2008.04
            Calc. date   ---> noviembre 12, 2015
==================================================================
```

Disk File Name = SRIM Outputs\Gold in aSiO2

Ion = Gold [79] , Mass = 196.967 amu

Target Density = 2.2100E+00 g/cm3 = 6.6451E+22 atoms/cm3
======= Target Composition ========
  Atom   Atom Atomic   Mass
  Name   Numb Percent Percent
  ----   ---- ------- -------
   Si   14   033.33   046.74
    O   8   066.67   053.26
=====================================
Bragg Correction = 0.00%
Stopping Units = keV / micron
See bottom of Table for other Stopping units

| Ion | dE/dx | dE/dx | Projected | Longitudinal | Lateral |
Energy	Elec.	Nuclear	Range	Straggling	Straggling
1.00 MeV	1.296E+03	2.929E+03	2580 A	357 A	320 A
1.10 MeV	1.331E+03	2.877E+03	2808 A	382 A	344 A
1.20 MeV	1.365E+03	2.826E+03	3037 A	407 A	367 A
1.30 MeV	1.398E+03	2.776E+03	3268 A	432 A	391 A
1.40 MeV	1.430E+03	2.728E+03	3499 A	457 A	414 A

Multiply Stopping by   for Stopping Units
--------------------   --------------------
 1.0000E-01   eV / Angstrom
 1.0000E+00   keV / micron
 1.0000E+00   MeV / mm
 4.5250E-03   keV / (ug/cm2)
 4.5250E-03   MeV / (mg/cm2)
 4.5250E+00   keV / (mg/cm2)
 1.5049E-01   eV / (1E15 atoms/cm2)
 1.1908E-04   L.S.S. reduced units
==
(C) 1984,1989,1992,1998,2008 by J.P. Biersack and J.F. Ziegler
```
```

Appendix C
Example of an Input File for TRIM

The following figure shows the input parameters used to obtain the ion distribution, the stopping powers as a function of the depth, and the ion range using the program TRIM. In this example the target material is amorphous silica and the impinging ions are 900 keV Au ions.

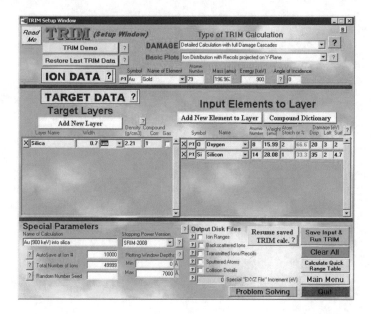

© Springer Nature Switzerland AG 2018
D. Bachiller Perea, *Ion-Irradiation-Induced Damage in Nuclear Materials*,
Springer Theses, https://doi.org/10.1007/978-3-030-00407-1

Appendix D
Example of an Input File for McChasy Code

```
 1  ***  MgO5e13                    (File name)
 2  TGT  MgO                        (Target material)
 3  KLM  001                        (Channeling Axis)
 4  ENE  1400                       (Proyectil energy expressed in keV)
 5  SCA  165                        (Scattering angle)
 6  FOR  trim                       (Program used to calculate dE/dx)
 7  COR  Mg 1.0       0.0     1400  (Stopping power correction for Mg)
 8  COR  O  1.0       0.05    1400  (Stopping power correction for O)
 9  BSP  on                         (Save backscattering spectra)
10  ESP  off                        (Save elemental spectra)
11  BEC  2.43 1                     (Channel width and offset)
12  OMC  0.41 1                     (Solid angle and cumulated charge)
13  RES  18                         (Energy resolution of the detector)
14  LEL  0.0                        (Local energy loss)
15  VIB  Mg 8.5                     (Atomic vibrations for Mg)
16  VIB  O  7.5                     (Atomic vibrations for O)
17  THI  2000                       (Sample thickness)
18  PAR  200000                     (Number of trajectories)
19  IBD  0.08                       (Ion-beam dispersion)
20  CRS  O O165.crs                 (Cross section file)
21  DEF  Mg 34  step  depth(nm)  fD (Defect profile)
22  1                 0         1.1
23  2                 10        2.0
24  3                 20        2.0
25  4                 40        2.7
26  5                 60        3.6
27  6                 80        4.0
28  7                 100       5.0
29  8                 120       5.6
30  9                 140       7.0
31  10                160       7.10
32  11                180       8.30
33  12                200       9.4
34  13                220       9.6
35  14                240       9.6
36  15                260       10
37  16                280       10
38  17                300       10.5
39  18                320       10.3
```

© Springer Nature Switzerland AG 2018
D. Bachiller Perea, *Ion-Irradiation-Induced Damage in Nuclear Materials*,
Springer Theses, https://doi.org/10.1007/978-3-030-00407-1

```
40   19                   340      10.1
41   20                   360      9.85
42   21                   380      10.4
43   22                   400      10.0
44   23                   425       9.7
45   24                   450       9.1
46   25                   475       8.7
47   26                   500       8.3
48   27                   550       8.0
49   28                   600      7.15
50   29                   650      6.00
51   30                   700      4.85
52   31                   800        0
53   32                   900        0
54   33                  1000        0
55   34                  1500        0
56   DEF O 34   step   depth(nm)    fD        (Defect profile)
57   1                     0       26.0
58   2                    10        2.0
59   3                    20        2.0
60   4                    40        2.7
61   5                    60        3.6
62   6                    80        4.0
63   7                   100        5.0
64   8                   120        5.6
65   9                   140        7.0
66   10                  160       7.10
67   11                  180       8.30
68   12                  200        9.4
69   13                  220        9.6
70   14                  240        9.6
71   15                  260        10
72   16                  280        10
73   17                  300       10.5
74   18                  320       10.3
75   19                  340       10.1
76   20                  360       9.85
77   21                  380       10.4
78   22                  400       10.0
79   23                  425        9.7
80   24                  450        9.1
81   25                  475        8.7
82   26                  500        8.3
83   27                  550        8.0
84   28                  600       7.15
85   29                  650       6.00
86   30                  700       4.85
87   31                  800         0
88   32                  900         0
89   33                 1000         0
90   34                 1500         0
91   GON static 0 0                          (Goniometer control)
92   EXE MgO5e13                             (Execute a task)
93   ***GON Rotate 4
94   ***EXE Random
95   ***AMO ON                               (Simulate amorphous spectrum)
96   ***EXE Amorphe
```

Printed in the United States
By Bookmasters